THE HAPTOGLOBIN GROUPS IN MAN

MONOGRAPHS IN HUMAN GENETICS

Vol. 4

BASEL (Switzerland) S. KARGER NEW YORK

THE HAPTOGLOBIN GROUPS IN MAN

by R. L. Kirk

John Curtin School of Medical Research
Australian National University, Canberra, A.C.T.

11 figures and 13 tables

1968

BASEL (Switzerland) S. KARGER NEW YORK

MONOGRAPHS IN HUMAN GENETICS

Vol. 1: L. BECKMAN (Uppsala): Isozyme Variations in Man.
VIII + 76 p., 33 fig., 22 tab., 1966. sFr. 19.80

Vol. 2: B. KJESSLER (Uppsala): Karyotype, Meiosis and Spermatogenesis in a Sample of Men Attending an Infertility Clinic.
VIII + 92 p., 19 fig., 15 tab., 1966. sFr. 23.50

Vol. 3: MORTON, NEWTON E.; CHUNG, CHIN S. and MI, MING-PI (Honolulu, Hawaii): Genetics of Interracial Crosses in Hawaii.
VIII + 160 p., 5 fig., 42 tab., 1967. sFr. 39.50

S. Karger AG, Arnold-Böcklin-Strasse 25, 4000 Basel 11 (Switzerland)

Printed in Switzerland by Buchdruckerei Effingerhof AG, Brugg

INDEX

1. THE COMMON HAPTOGLOBIN GROUPS

Early in 1955 OLIVER SMITHIES [188, 189], using a new technique of electrophoresis in a gel made from partially hydrolysed potato starch, demonstrated that sera from normal adults could be divided into three groups. SMITHIES examined sera from only 43 individuals but he was able to demonstrate 10 persons belonging to what he called group I, 21 to group II A and 12 to II B. These groups appeared to be independent of the AB0 or Rh blood types of the persons concerned and there was no significant difference in distribution between males and females.

Collaborating with NORMA FORD WALKER, SMITHIES [197] followed up his first observation with a study of parents and children in 18 families. They postulated that the three groups in the population were under the control of a single pair of genes and their family study supported this hypothesis. In the following year SMITHIES and WALKER concluded that the serum protein which gave rise to the differences observed after electrophoresis in starch gel was identical with an α_2, haemoglobin-binding protein which had been studied extensively since 1938 by French workers [43, 89, 103–106, 108, 134, 164] and others and which had been called haptoglobin. Accordingly SMITHIES and WALKER proposed the terminology for the haptoglobin group system which is now in use, and further confirmed their postulated mode of inheritance based on a pair of co-dominant autosomal alleles Hp^1 and Hp^2.

Table I. The common haptoglobin groups

Original notation	Present notation	Genotypes
Group I	Hp 1–1	Hp^1/Hp^1
Group IIa	Hp 2–1	Hp^1/Hp^2
Group IIb	Hp 2–2	Hp^2/Hp^2

The three common haptoglobin phenotypes listed above have been detected so far in all populations studied, though their frequency varies considerably in different parts of the world. In addition black African and American Negro populations show an appreciable frequency of a modified Hp 2–1 pattern and also a proportion of persons with no detectable haptoglobin pattern at all after normal starch-gel electrophoresis. This latter phenotype, Hp 0, is assigned to persons who are said to be ahaptoglobinaemic, or perhaps more correctly, hypo-haptoglobinaemic. Persons giving an Hp 0 phenotype are found in many parts of the world in low frequency and the significance of this will be discussed more fully in section 4 below.

When serum from normal persons is electrophoresed in starch-gel characteristic patterns can be observed [191]. Since all the haptoglobin components bind with haemoglobin the patterns may be visualised more easily by using the peroxidase reaction of haemoglobin to bring about colour reactions with benzidine [134], o-tolidine [4], leucomalachite green [124], guaiacol [169], and o-dianisidine [157].

Sera from Hp 1–1 persons give only a single, intensely staining band somewhat slower than free haemoglobin at pH 8.6 in borate buffer. Hp 2–2 persons give a pattern in which there is no band in the position of the Hp 1–1 component, but a series of slower bands of diminishing intensity towards the origin. Sera from Hp 2–1 persons,

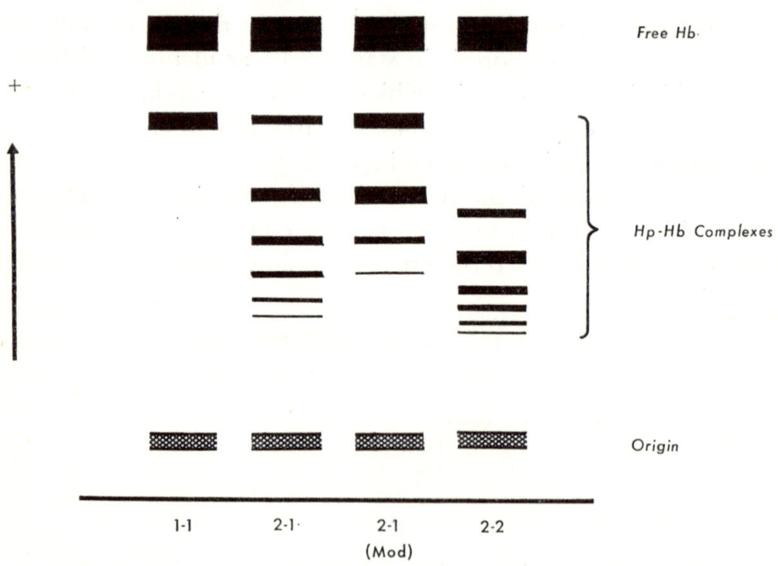

Fig. 1. The common haptoglobin phenotypes. Starch-gel electrophoresis of the Hp-Hb complexes at pH 8.6.

who are genetically heterozygous, give a pattern similar to that for Hp 2–2, but in which the bands have a slighter greater mobility which displaces them toward the anode, together with a band corresponding to the Hp 1–1 position. The significance of these patterns, which are illustrated in Figure 1, will be discussed fully in section 5 below, but they correspond to the existence of a series of polymer associations of the haptoglobin molecules, the larger-sized particles moving more slowly through the pores of the starch-gel under the influence of the potential gradient along the gel.

The Hp 2–1 (modified) phenotype which is relatively common in black Africans and American Negroes, gives a pattern in which the slower bands of the Hp 2–1 pattern are absent, and in which there is an increased concentration of the fastest bands. Inexperienced observers sometimes classify Hp 2–1 (mod) as Hp 2–1.

A. The Sub-Types of Haptoglobin

Six years after the discovery of the haptoglobin groups Smithies and his collaborators made an equally important advance announced at the 2nd International Congress of Human Genetics in Rome in 1961 [50, 193]. This has led to a more significant understanding of the relationship between the Hp genes and the protein structures they control. If the haptoglobin (free from haemoglobin) in serum is partially purified on a DEAE cellulose column and then subjected to reduction with mercaptoethanol in a concentrated urea solution the bonds holding together the polypeptide units of the haptoglobin molecule become broken. Electrophoresis in a starch-gel at acid pH and containing 8M urea can then be used to separate the polypeptides thus produced.

The application of this elegant technique by Smithies and his collaborators revealed that haptoglobin consists of two polypeptide chains, α and β. The β-chains are apparently the same in all persons (with the exception of the Hp-Marburg and Hp-Bellevue phenotypes discussed below) and are therefore not affected by the majority of mutations in the Hp gene. In the common phenotypes, however, three different α-chains have been noted which vary in their mobility and intensity of staining after electrophoresis in 8M urea starch gel.

Persons with phenotype Hp 1–1 can give rise to a fast migrating α-polypeptide (hp 1F), to a slightly slower migrating α-polypeptide (hp 1S) or to a mixture of both hp (1F+1S). Hp 2–2 persons have an α-polypeptide band on the gel, hp 2, considerably slower than either hp 1F or hp 1S. Hp 2–1 persons may have the hp 2 band combined with either hp 1F or hp 1S. The Hp 2–1 (mod) phenotype shows no difference in the hp 1F or hp 1S bands but reveals a characteristic reduction in the intensity of the hp 2 band. This α-polypeptide is called hp 2m.

The existence of two variants of the hp 1 α-polypeptide suggests that there are two Hp^1 alleles, Hp^{1F} and Hp^{1S}. Family studies carried out by Smithies and his collaborators [195], and by Goedde *et al.* [81] support this concept, and the main phenotypes and genotypic combinations are given in Table II.

Table II. The sub-types of haptoglobin

Phenotypes at pH 8.6	Phenotypes after reductive cleavage	Genotypes	α-peptide band patterns in 8M urea starch-gel 2 α	1S α	1F α
1–1	1F–1F	Hp^{1F}/Hp^{1F}			■
	1S–1S	Hp^{1S}/Hp^{1S}		▮	▮
	1F–1S	Hp^{1F}/Hp^{1S}		■	
2–1	2–1F	Hp^2/Hp^{1F}	▮		▮
	2–1S	Hp^2/Hp^{1S}	▮	▮	
2–2	2–2	Hp^2/Hp^2	■		

Reductive cleavage followed by electrophoresis in 8M urea starch gel requires care to achieve reproducible results. Its application, however, increases the number of recognizable common haptoglobin phenotypes (and genotypes) in most populations to six, and to eight phenotypes for black African and American Negro populations. It should be noted that the ninth possible genotype Hp^2/Hp^{2M} cannot be detected phenotypically as different from Hp^2/Hp^2 and the tenth genotype Hp^{2M}/Hp^{2M} is probably Hp 0.

B. A Case of Unequal Crossing-Over

When the α-polypeptides from the homozygous Hp^{1F}/Hp^{1F} and Hp^{1S}/Hp^{1S} persons were broken down into smaller peptides by chymotryptic digestion, Smithies, Connell and Dixon [194] found the 'finger prints' obtained were identical except for one peptide. Complete amino-acid analysis of the hp 1S α-polypeptide showed that it differed from the hp 1F α-polypeptide by the replacement of a single lysine residue in the 1F α-polypeptide by an acidic amino acid in the 1S α-polypeptide. The resulting charge difference, of course, accounts for the difference in mobility of these two polypeptides when subjected to electrophoresis in 8M urea starch gel. It explains also the difference in the single peptide of the 'finger print'.

Examination of the 'finger print' of peptides obtained by chymotryptic digestion from the hp 2 α-polypeptide revealed that all the peptides produced by digestion of hp 1S and hp 1F were present together with one additional peptide. Further, two of the peptides which included the terminal sequences of one end of each of the hp 1S and hp 1F polypeptides, although present in the 'finger print' from the hp 2 α-polypeptide, were reduced in intensity, and the new peptide after amino-acid analysis was found to be slightly smaller than that which would be obtained by joining together the two terminal peptides showing reduced intensity. Finally, the molecular weight of the hp 2 α-polypeptide is approximately twice the molecular weight of either the hp 1F or hp 1S α-polypeptides.

These observations led Smithies, Connell and Dixon [194] to advance the hypothesis that the *Hp*^{2} gene arose as a result of non-homologous chromosomal crossing-over involving the cistrons for the hp 1F and the hp 1S α-polypeptides in a person heterozygous for these genes. Diagrammatically, the proposal is as shown below (Fig. 2).

Non-homologous crossing-over between the chromatids carrying the cistrons for the Hp 1F and the hp 1S α-polypeptides resulted in a new chromosome containing a partial 'duplication' or a cistron almost twice as long as either of the hp 1 cistrons, together with a chromosome carrying a 'deleted' section which was presumably non-functional in producing haptoglobin protein. The α-polypeptide synthesized as a result of the message coded on the 'duplicated' chromosome will contain one peptide corresponding to the N-terminal sequence of the hp 1F α-polypeptide and one peptide corresponding to the C-terminal sequence of the hp 1S α-polypeptide. A portion of the N-terminal sequence of hp 1S and of the C-terminal sequence of the hp 1F α-polypeptides will have been joined together to form what appears on the 'finger print' of the hp 2 α-polypeptide as a completely new peptide spot. This will be nearly, but not quite the size of the N and C terminal peptides joined together.

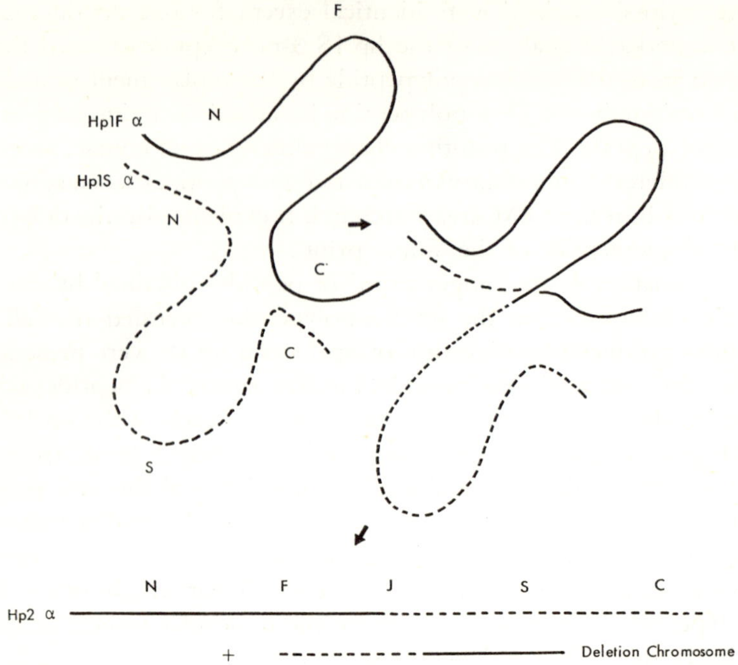

Fig. 2. Diagramatic representation of the evolution of the Hp^2 gene by non-homologous crossing-over. Based on Smithies *et al.* [194].

The correlation of cytological demonstration and phenotypic effect of unequal crossing-over leading to gene-duplication was made first by C. B. Bridges [44] in the case of the Bar-eye phenomenon in the fruit-fly *Drosophila melanogaster*, although the cytological mechanism by which it could occur had been demonstrated by Serebrovsky [183] some years earlier. Further, gene-duplication has been invoked by Lewis [126] as an important mechanism in the evolution of genetic systems, and this concept has been used by Ingram and Stretton [99] to explain the possible evolution of the genetic loci controlling the synthesis of the haemoglobin polypeptides in man. Smithies, Connell and Dixon [192], however were the first workers to produce evidence of the peptide and amino-acid changes which could be expected from an almost, but not quite complete duplication, and the further extension of this mechanism to explain the occurrence of some of the rarer haptoglobin phenotypes will be outlined in the next section.

From a consideration of the probabilities involved in producing a cross-over event of this kind SMITHIES [192] has suggested more recently that the formation of the hp 2 cistron was a unique occurrence at some time in the past. The altered cistron has now become established and is universally distributed in populations in all parts of the world. He has postulated that the Hp^2 gene might be favoured by selection in the homozygous state (see p. 52) but that it will not completely replace the Hp^1 gene. This is because new Hp^1 genes will be added to the gene pool from time-to-time as a result of unequal crossing-over following displaced synapsis in persons homozygous for the Hp^2 gene. A similar argument had been advanced previously by NANCE [148].

2. RARE HAPTOGLOBIN GROUPS

A considerable number of rare haptoglobin phenotypes have now been described (Table III). These may best be understood if the following facts are remembered.

The haptoglobin molecule consists basically of two polypeptides α and β, and the structure of the α-polypeptide is under the control of three genes Hp^{1F}, Hp^{1S} and Hp^{2}. The hp 2 α-polypeptide is approximately twice the size of either of the hp 1 α-polypeptides, and in addition it normally has a strong tendency to polymerise so that the Hp 2–2 phenotype gives a series of bands each with decreasing mobility as the molecular weight increases. Correspondingly, the Hp 2–1 phenotype also shows a series of bands consisting of association of hp 1 and hp 2 and polymers of hp 2. In addition some non-associated hp 1 is present in the position of the single band of the Hp 1–1 phenotype. Finally, haptoglobin binds haemoglobin, and the mobility in starch-gel of the Hp-Hb complex is lower than of haptoglobin alone.

Mutations may affect the haptoglobin phenotypes in several possible ways.

1. Alteration in electric charge of either the α or β polypeptides by an appropriate amino-acid substitution. Such a change satisfactorily accounts for the difference between the hp 1F and hp 1S polypeptides. A charge difference of this kind may be masked, however, when the α-polypeptide is associated with the β-polypeptide. Thus no difference can be detected on starch gel in the intact hp 1 molecule derived from a homozygous Hp^{1F}/Hp^{1F} or a Hp^{1S}/Hp^{1S} person before reductive cleavage has been carried out.

2. Alteration in the rate of synthesis of either the α or β-polypeptides, resulting in a charge in the amount of the intact molecules detectable. Such mutations are known for the α and β-polypeptides of human haemoglobin, the deficiencies giving rise to the so-called α and β-thalassaemias.

3. Chromosomal rearrangements. As pointed out above there is strong evidence that partial duplication of the hp 1F and hp 1S α-

Table III. Rare phenotypes in the haptoglobin system

Abnormal phenotypes observed in starch gel at pH 8.6	Notes	Suggested gene symbol	Genotypic combinations with normal phenotypes	Genotypic combinations with abnormal phenotypes
Hp 2–1 'Carlberg'	Reduced hp 1 synthesis	Hp^{1ca}	Hp^{1ca}/Hp^{1}	Hp^{1ca}/Hp^{1ca} Hp^{1ca}/Hp^{2}
Hp 2–1 'Johnson'	a 'Triplicated hp α chain	Hp^{J}		Hp^{1}/Hp^{J} Hp^{2}/Hp^{J} Hp^{J}/Hp^{J} } Reduced synthesis
Hp 1–P Hp 2–P	Observed only on normal gel when complexed with Hb	Hp^{P}		Hp^{P}/Hp^{P}, Hp^{P}/Hp^{1S} Hp^{P}/Hp^{1F}, Hp^{P}/Hp^{2}
Hp 2–L	Observed only on normal gel when complexed with Hb	Hp^{L}		Hp^{L}/Hp^{2}, Hp^{L}/Hp^{1} ? Hp^{L}/Hp^{L} ?
Hp 1–H	Observed on normal gel both complexed and non complexed with Hb	Hp^{H}		Hp^{H}/Hp^{1}, Hp^{H}/Hp^{2} Hp^{H}/Hp^{H} ?
Hp 2–1 (Trans)	Reduced hp 2 synthesis	Hp^{2Tr}	Hp^{2}/Hp^{2Tr}	Hp^{1}/Hp^{2Tr} Hp^{2Tr}/Hp^{2Tr} (Hp 0 ?)
Hp 2–1 (Haw)	Reduced hp 2 synthesis	Hp^{2Haw}	Hp^{2}/Hp^{2Haw}	Hp^{1}/Hp^{2Haw} Hp^{2Haw}/Hp^{2Haw} (Hp 0 ?)
Hp Ab	Reduced hp 2 synthesis	Hp^{Ab}	Hp^{2}/Hp^{Ab}	Hp^{1}/Hp^{Ab} Hp^{Ab}/Hp^{Ab} (Hp 0 ?)
Hp 2–1D	Not detectable when complexed with Hb	Hp^{1D}		Hp^{1D}/Hp^{1D}, Hp^{1S}/Hp^{1D} Hp^{1F}/Hp^{1D}, Hp^{2}/Hp^{1D}
Hp 1–B Hp 2–B	possibly a 1 α- or a 2 α-chain mutation	Hp^{B}		Hp^{1}/Hp^{B}, Hp^{2}/Hp^{B} Hp^{B}/Hp^{B}
Hp 1-'Marburg' Hp 2-'Marburg'	An hp β-chain mutation	Hp^{Mb}		Hp^{Mb}/Hp^{Mb}, Hp^{1}/Hp^{Mb} Hp^{2}/Hp^{Mb}
Hp 2–1 'Bellevue'	A hp β-chain mutation	Hp^{Bell}		Hp^{1}/Hp^{Bell}, Hp^{2}/Hp^{Bell} Hp^{Bell}/Hp^{Bell}

polypeptide cistrons is a highly probable explanation of the origin of the Hp^2 gene. Other chromosomal rearrangements might also be expected, and as will be noted below, appear to have occurred.

Hp 2–1 'Carlberg'

The first rare haptoglobin phenotype was described by GALATIUS-JENSEN [68] in 1958. It was found in Denmark in a mother referred for examination in a paternity case, and a similar phenotype was detected later in another Danish male, his brother and a son. The variant was called Hp 'Carlberg' after the family in which it was first detected.

Hp 2–1 'Carlberg' can easily be typed as Hp 2–2 in normal starch gel, though close examination reveals additional bands which apparently correspond to the fastest moving bands in the Hp 2–1 pattern (Fig. 3).

Further examples of Hp 2–1 'Carlberg' have been described by HARRIS *et al.* [91], and GIBLETT [74] and by BLOOM *et al.* [39]. GIBLETT has subjected some samples of this variant to reductive cleavage, and after electrophoresis in 8M urea starch gel she found bands corresponding to the hp 1S α-polypeptide and the hp 2 α-polypeptide. The hp 1S, however, was greatly reduced in intensity, suggesting that Hp 2–1 'Carlberg' is caused by a mutant Hp^{Ca} gene which decreases

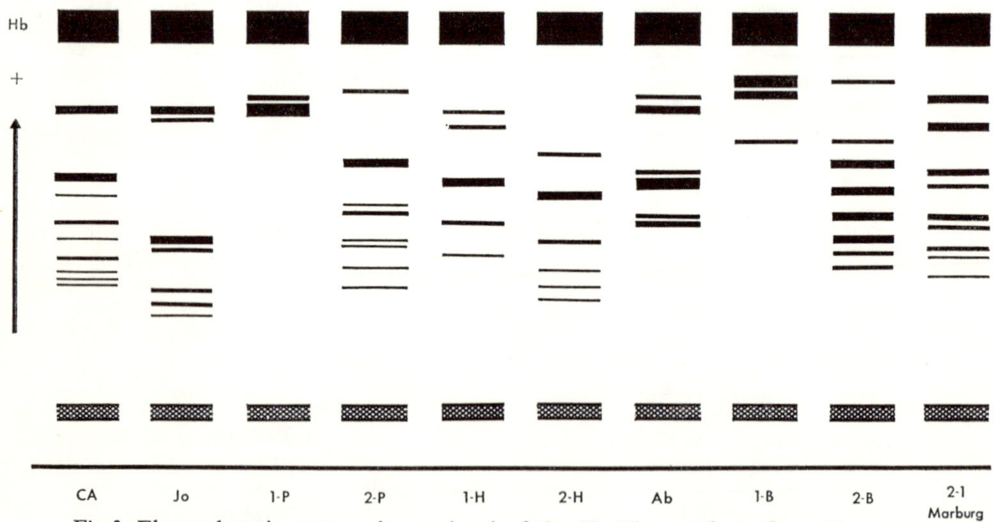

Fig. 3. Electrophoretic patterns in starch-gel of the Hp-Hb complexes from 10 rare phenotypes in the haptoglobin system. Relative mobilities are the same as those in Fig. 1.

the amount of hp 1, leaving an excess of hp 2 in the heterozygote to appear as bands in the normal Hp 2–2 position.

BERNINI *et al.* [33] have reported that a phenotype corresponding to that of the 'Carlberg' phenotype can be reproduced by mixing hp 1 α and hp 2 α-polypeptides in the proportion of 1:9, supporting the interpretation that this phenotype is due to a mutant gene suppressing hp 1 α production.

Hp 2–1 'Johnson'

This was the second rare haptoglobin phenotype to be described. Detected first of all in the serum of an American Negress and her daughter [77, 195], additional examples have been found in Europeans from Australia, Denmark, Holland Norway, Sweden, Poland and in two Hawaiian Chinese, in a Kurdish Jew in Israel and an Australian Aboriginal [55, 171, 179, 194]. The Hp 2–1 Johnson phenotype is characterised in normal starch-gel by a double band in the position of the Hp 1 band, and another double band and a series of bands of decreasing intensity in a very retarded position on the gel. Subtyping after reductive cleavage shows a normal hp 1S α-polypeptide and an hp α-polypeptide considerably slower than the normal hp 2 α [74]. SMITHIES, CONNELL and DIXON [194] have suggested that this new hp α-polypeptide is the product of an Hp^J gene which could have arisen by unequal crossing-over between the chromatids of a homozygous Hp 2–2 individual to yield a cistron which is partially triplicated (Fig. 4). (The increased molecular weight of the new polypeptide controlled by the hpJα cistron accounts for its extremely slow migration in the gel.)

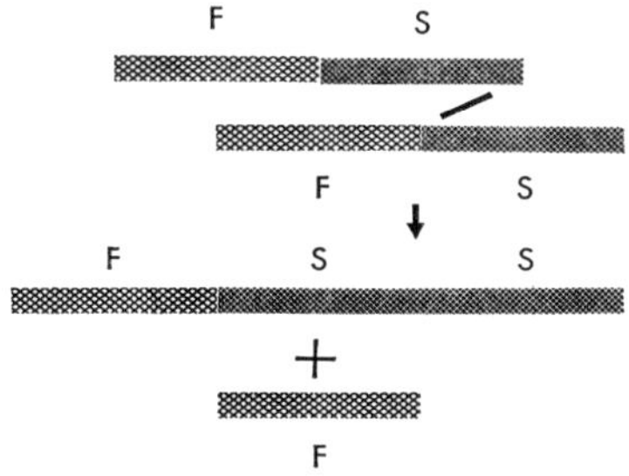

Fig. 4. Possible mechanism for the production of the 'Johnson'-type cistron by unequal crossing-over between two Hp^2 cistrons.

SMITHIES and his colleagues point out that some of the 'Johnson' phenotypes sent to them for testing are not all precisely identical, and they suggest that this may not be surprising if the hp J cistron arises in the way postulated. The result of such an unequal crossing-over will depend on the type of the hp 2 α cistron involved, and they point out that theoretically one may expect four such hp 2 α cistrons, and that therefore eight possible Hp J cistrons may also be expected. The occurrence of the four types of hp 2 α cistrons will be discussed further below.

Two variants in a single sibship were described as 'modified' Johnson types by GOHLER and BUNDSCHUH [82]. The propositus was suffering from tuberculosis, but his brother who showed a similar variation in the normal pattern was in good health. As in the typical 'Johnson' pattern, there is a wide separation between the band of greatest mobility and the next slower band.

Hp 1–P and Hp 2–P

ROBSON *et al.* [175] in 1964 described two unusual phenotypes found in members of a British family. There were altogether six Hp 1–P and two Hp 2–P persons in a single pedigree. The unusual patterns were present only when electrophoresis in starch-gel at pH 8.6 was carried out using serum to which haemoglobin had been added. The patterns are shown in Figure 3. When partially purified, uncomplexed haptoglobin was used the patterns were indistinguishable from normal Hp 1–1 or Hp 2–1 phenotypes. Subtyping of the Hp 1–P persons revealed an Hp 1 band indistinguishable from that of a normal hp 1S α-polypeptide. Similarly subtyping of the Hp 2–P type gave two bands indistinguishable from the hp 1S and hp 2 α-polypeptides. Since there are no slow-moving bands in the Hp 1–P pattern ROBSON and her colleagues suggested that the variant gene Hp^{P} is a modification of the Hp^{1} gene.

Hp 2–L

Individuals with another unusual pattern were also observed by the previous authors in one family of mixed Creole and Chinese descent from the West Indies. The pattern in normal starch-gel (Fig. 3) of the complexed haptoglobin resembles that of Hp 2–2 but each of the bands has a slightly increased mobility. Partially purified uncomplexed Hp 2–L gives a pattern indistinguishable from normal Hp 2–2. On sub-typing only a normal hp 2 α-polypeptide band could be distinguished.

Both the P and L variants are similar in showing differences only when complexed with haemoglobin, with no variation detectable in the respective 1 or 2 α-polypeptides. The fact that the Hp 1–P phenotype has no slow moving bands suggests that the mutant gene is a variant of Hp 1S, whilst the Hp 2–L phenotype so closely resembles that of a normal Hp 2–2 that it suggests the mutant is a variant of Hp^2.

If the Hp^P and Hp^L genes control structural alterations in the haptoglobin molecule it seems reasonable to suppose that this in some way is exposed by the binding of haemoglobin since no differences are detected in the uncomplexed intact molecule or in the separated α-polypeptides.

Shim, Lee and Kang [186] however, suggest that the P and L variants are due to mutations affecting one of the haemoglobin binding sites on the β-polypeptide chain of haptoglobin. Further studies will be needed to determine the precise nature of the changes involved in these cases.

Robson and her colleagues pointed out if that the P mutation affected the Hp S α-polypeptide and the L mutation the Hp 2 α-polypeptide, the original structural change could have been the same. The mutation of the Hp S cistrons could then have become incorporated into an Hp 2 cistron by unequal crossing-over in an appropriate heterozygote. Alternatively the Hp^L mutation could have arisen independently and in view of the fact that it was found in a family of mixed Creole and Chinese ancestry, whereas the Hp^P mutation was observed in a British family, this seems the more plausible explanation.

In passing it should be noted that the Hp 2–L phenotype may be more frequent than is apparent so far. It is so similar to Hp 2–2 that under normal typing conditions it may readily be misclassified, even by experienced observers.

Hp 1–H and Hp 2–H

Finally, Robson and her colleagues found two members of another British family showing a variant phenotype, one in connection with an Hp^1 allele, the other with an Hp^2 allele. When complexed with Hb the Hp 1–H pattern resembles that of a Hp 2–1 person with an additional band just behind the Hp 1 position but the slower moving bands of the Hp 2–1 pattern are missing. Partially purified uncomplexed haptoglobin from this person still showed the additional band behind the Hp 1 band. The Hp 2–H pattern resembled that of the Hp 2–2 person, with the addition of an extra band midway between the first and second bands of the normal Hp 2–1 pattern.

The appearance of the bands suggested to the authors that Hp 1–H and Hp 2–H are due to an *Hp*^{2} mutation, in combination respectively with normal *Hp*^{1} or *Hp*^{2} alleles. They reported also that Dr. W. NANCE had found a similar phenotype in a Brazilian Indian, and that on sub-typing his serum only a normal Hp 2 α-polypeptide band was found.

Hp 2–1 (trans) or Hp 2–1 (Int)

GIBLETT in 1964 [74] summarized what was known at that time about rare haptoglobin phenotypes and added three more, Hp 2–1 (trans), Hp 2–1 (Haw) and Hp Ab.

The Hp 2–1 (trans) phenotype has been found in non-Negro persons, mainly those of Oriental origin. It is characterized by a pattern of the 2–1 type but with a change in the concentration of the bands similar to, but not identical with that found in the Hp 2–1 (mod) phenotype. The two faster moving bands are more intense, as is true also for Hp 2–1 (mod), than in the normal Hp 2–1 pattern, but it is distinguished from Hp 2–1 (mod) by the greater intensity of bands 3 and 4. Sub-typing reveals no difference in mobility of the hp F, S or 2 α-polypeptides, but the ratio of the hp 2 α-bands to the F or Sα-band is consistently reduced when compared with a normal Hp 2–1. This suggests that the mutation is affecting the rate of production of the *Hp*^{2} gene product in a way similar but not identical with that for the *Hp*^{2m} mutation.

Hp 2–1 (Haw)

One example of this variant was found by GIBLETT among 1300 samples from Hawaii. It is not stated what was the ethnic background of the propositus, but the phenotype in normal starch-gel again resembles that of Hp 2–1 (mod). There is a marked increase in concentration of the hp 1 band, but there is not a corresponding increase in the intensity of the second band nor decrease in intensity of the slower bands when compared with a normal Hp 2–1 pattern. On sub-typing the pattern is apparently identical with that of an Hp 2–1 (mod) pattern. It seems likely that the Hp 2–1 (Haw) phenotype represents an *Hp*^{2} gene mutation in combination with a normal *Hp*^{1} allele. The *Hp*^{2} (Haw) mutant apparently depresses the production of the hp 2 α-polypeptide, but allows more of the higher polymers to be formed than in the case of the *Hp*^{2m} gene.

Hp Ab

GIBLETT reports also an unusual phenotype found by Dr. IRVING UMANSKY in the serum of a Boston woman. The major bands in this pattern correspond with those usually present in the Hp 2–1 (mod) pattern, but with at least one additional minor component moving ahead of the major bands. The sub-typing of Hp Ab closely resembles that of an Hp 2–1 (mod) pattern. This again suggests an alteration in the Hp^2 gene product, but it is difficult to explain the appearance of the minor components. Possibly there is not only a change in the rate of production of the hp 2 α-polypeptide, but an alteration in the binding capacity with haemoglobin. Further studies of this phenotype are clearly necessary.

Hp 2–1D

Recently, RENWICK and MARSHALL [173] have described a British family in which two examples of an unusual Hp phenotype were observed. In contrast to the P and L variants noted above the Hp 2–1D phenotype is indistinguishable from a normal Hp 2–1 phenotype when complexed with haemoglobin. However, in starch-gel at pH 8.6 the hp 1 band of the uncomplexed haptoglobin migrates slightly faster than the normal hp 1 band. Sub-typing revealed an hp 1 α-polypeptide of slightly greater mobility than the hp F α-polypeptide.

The results suggest a structural hp 1 α-mutation. The charge difference does not seem fully expressed, and when complexed with haemoglobin it is completely masked.

Hp 1–B and Hp 2–B

GIBLETT, UCHIDA and BROOKS [80] have now reported another two rare phenotypes which show distinctive patterns in normal starch-gel and on sub-typing. The new types were found in a family which was being examined because of a chromosomal anomaly detected in one member. The abnormal chromosome anomaly, however, was not associated with the unusual haptoglobin phenotypes. The Hp 1–B pattern consists of three bands, two with rates of migration close to those for Hp 1 and the third somewhat retarded (Fig. 3). Hp 2–B has two faint bands corresponding to the 1st and 3rd bands of Hp 1–B followed by a series of fairly evenly spaced bands which appear to be characteristic of this phenotype. When not complexed with haemoglobin the haptoglobin patterns of Hp 1–B and Hp 2–B were still readily distinguishable from the normal patterns.

On sub-typing the Hp 1–B showed in addition to a normal hp 1 α-band an hp B-band with a migration rate intermediate between hp IS α and hp 2 α. Similarly, Hp 2–B showed a normal hp 2 α-band and the new hp B.

Since the new polypeptide migrates at a rate intermediate between hp IS α and hp 2 α the *Hp*^{B} gene could be a structural mutation of either *Hp*^{1} or *Hp*^{2}. Alternatively, *Hp*^{B} might be conceivably the result of a partial duplication by involving different lengths of the IF and IS cistrons than that which gave rise to the normal *Hp*^{2} allele. Only future study of the 'finger print' and amino-acid sequences of the hp B-polypeptide will resolve these possibilities.

Hp 'Marburg'

Another recent discovery is that of a possible Hp β chain variant. CLEVE and DEICHER [49, 211] found three examples in a German family showing two abnormal phenotypes, Hp 1 'Marburg' and Hp 2 'Marburg'. In starch-gel at pH 8.6 the Hp 1 'Marburg' is characterised by a double band in the hp 1 position, and an additional faint band migrating slightly more slowly. In the Hp 2 'Marburg' the hp 1 band is split as before, and in addition the 2nd and 3rd bands of the normal Hp 2–1 pattern are split also.

Sub-typing revealed no detectable difference in the hp 1 α or hp 2 α-polypeptides, but consistent though slight differences were noted in the pattern of the hp β-polypeptides in 8M urea starch gel. Haptoglobin 'Marburg' thus becomes the first example of a probable mutation in the cistron for the Hp β-polypeptide. If this finding is confirmed, further examination of this family may yield valuable information on the linkage relationship between the cistrons for the hp α- and β-polypeptides.

Hp 2–1 'Bellevue'

Another β-chain alteration has been observed in the serum of an American Negro male by JAVID [102]. In the Hb-complexed state after starch-gel or acrylamide-gel electrophoresis the pattern shows a fast anodal band identical with that in a normal Hp 2–1 phenotype, but the remaining bands are split and do not correspond to the mobilities of the other bands present in the normal Hp 2–1 pattern. Similar changes in pattern were noted for the purified haptoglobin both complexed with Hb and in the free state.

Reductive cleavage revealed a normal hp α band, but an additional faster component in the hp β-polypeptide.

Examination of the children of the propositus was difficult because of extremely low haptoglobin levels in the serum. The evidence suggested, however, that one child was probably Hp 1–1 'Bellevue' and another Hp 2–2 'Bellevue'.

JAVID, in reporting the Hp 2–1 'Bellevue' phenotype points out that the hp β chain locus has not been given a name, and since haemoglobin is bound to the β-polypeptide, he suggests the symbol B_p for 'Binding peptide'.

The suggestion is not a fortunate one, since it introduces a completely new symbol with little mnemonic significance. If two Hp locus symbols are required Hp_{α} and Hp_{β} would be more appropriate. The genotypic designation of the Hp 2–1 'Bellevue' individual would then be:

$Hp^{1F}_{\alpha}/Hp^{2}_{\alpha} : Hp^{A}_{\beta}/Hp^{Bell}_{\beta}$

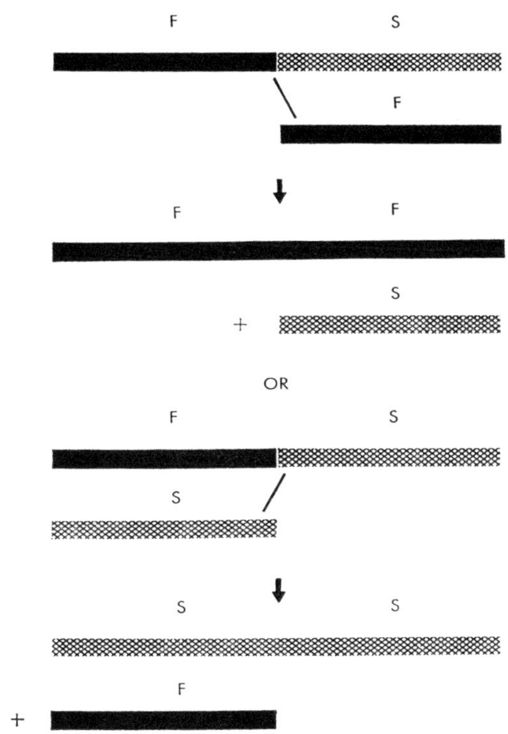

Fig. 5. Possible mechanism for the production of unusual Hp^2 cistrons by crossing-over between 'normal' Hp^2 and Hp^{1F} or Hp^{1S} cistrons.

Variants Detected only by Sub-typing

The brilliant suggestion by Smithies, Connell and Dixon [194] that the normal Hp 2 α-cistron was the result of a partial duplication of the Hp IS and Hp IF α-cistrons in a heterozygous Hp^{1F}/Hp^{1S} individual has been discussed in section 1. At that time Smithies and his colleagues pointed out that if this explanation was correct other Hp 2 α-cistrons could be formed in the following manner by crossing over within the α-haptoglobin locus in persons heterozygous for Hp^1 and Hp^2 (Fig. 5).

It was thought probable that since the S and F chains alone migrate at different speeds in 8M urea gel, the predicted hp 2 α-chains produced by the cistrons Hp 2FF and Hp 2SS would migrate, under the same conditions, at speeds different from the 2FS (or 2SF) chains produced by the usual Hp^2 gene.

Nance and Smithies [149] reported in 1963 what appeared to be a striking confirmation of this prediction. In 12 out of 500 Brazilian families they found unusual Hp^2 alleles on sub-typing in 8M urea starch gel, though the Hp phenotypes at pH 8.6 were always identical with those expected for a normal Hp^2 allele.

In all the cases studied two types of unusual Hp 2 α-polypeptide bands were observed, one migrating 3% faster and the other 1.5% slower than the normal band. This is consistent with the idea that the faster one represents an 'FF' polypeptide and the slower one an 'SS' polypeptide. It is possible that similar changes could be produced by appropriate single amino-acid substitutions in the 2 α-polypeptide. Final proof of the correctness of the hypothesis must await complete 'Finger-print' and amino-acid analysis of the new variants.

3. THE FORMAL GENETICS OF THE HAPTOGLOBIN GROUPS

A. The Two-Allele Model

Since sub-typing is a relatively expensive and difficult procedure most studies of the inheritance and distribution of the haptoglobin groups have provided data for the types detectable in normal starch gels at pH 8.6. For those populations where the three common phenotypes are Hp 1–1, 2–1 and 2–2 an autosomal two-allele model Hp^1 and Hp^2 appears to be adequate to explain the inheritance, as suggested originally by SMITHIES and WALKER [197].

In family studies the expected distribution of the Hp groups among children should be:

Table IV. Expected distribution of children in various mating classes

Phenotypes of parents	Expected proportion among children of 1–1	2–1	2–2
1–1 X 1–1	1.0	0	0
1–1 X 2–1	0.5	0.5	0
1–1 X 2–2	0	1.0	0
2–1 X 2–1	0.25	0.5	0.25
2–1 X 2–2	0	0.5	0.5
2–2 X 2–2	0	0	1.0

Close approximation to these expectations should be achieved if there is no selection of gametes before fertilization (pre-zygotic selection) or selection against particular genotypes in the foetal or neonatal period (post-zygotic selection). Further complications could arise if the ability to classify the Hp phenotype correctly varies with age. GALATIUS-JENSEN [69] pointed out that new-born babies are difficult to classify, some 90% of cord-blood samples appearing as Hp 0. At the age of 1 month this proportion had dropped to 50%, and above the age of four months he found only rare cases where it was not possible to assign an Hp group to the child. BECKMAN and GRIVEA [24] have re-examined the problem of classifying new-born children: they were able to classify 23% of cord-blood samples and 64% of serum samples from children aged 1 week. The inability to classify the Hp groups of young children, however, does not mean that they are wrongly classified. On the contrary if children aged less than six months are excluded from family studies no individual should be incorrectly classified and only rarely should either adults or children in normal good health be classified as Hp 0. The problem of the rare Hp 0 persons in European populations and of their occurrence in black African or American Negro populations will be discussed more fully in section 4 below.

a. Family Studies

Sixteen published studies include more than 2,000 families with 5,000 children (Table V). Thirteen of these were concerned with persons of European origin; that of STEINBERG *et al.* [201] with Chinese and Indian families in Malaysia, and those of STEINBERG and MATSUMOTO [201a] and of MATSUNAGA *et al.* [132] with Japanese. The distribution of children by phenotype for each of the parental mating classes is shown in Table Va.

With few exceptions, which are dealt with more fully below, the haptoglobin groups of the offspring agree with what is expected on the basis of a two-allele model. Omitting the atypical segregants and the Hp 0 offspring, the deviations from expectation for any particular mating class are relatively small.

GALATIUS-JENSEN [67] pointed out that in the family studies published up to that time and including his own that there is a tendency for an excess of heterozygous children to be born to parents of the mating class Hp 2–1 by Hp 2–2. HARRIS, ROBSON and SINISCALCO [91] added further information for this mating class, and the deviation from expectation for the combined figures at that time was significant. The results in Table Va based on 16 published studies show a deviation in the same direction, though its magnitude is not as great as that reported in the earlier investigations. Further, if the Hp 0 children are mainly of phenotype Hp 2–2, their inclusion in the total would reduce the discrepancy even further.

More recently MORTON, KRIEGER and MI [135] have analyzed more than 1,000 Brazilian families for segregation distortions. The results

Table V. Family studies involving the three common haptoglobin phenotypes

Authors	No. of families	No. of children	Ref. No.
BAITSCH *et al.*	353	686	14
FALK and PROKOP[2]	217	466	61
FLEISCHER and MOHR[1]	126	428	64
FLEISCHER and LUNDEVALL	93	296	63
GALATIUS-JENSEN	205	509	67
HARRIS *et al.*[1]	300	788	91
HERZOG *et al.*	33	61	95
MÄKELÄ *et al.*	126	419	128
MATSUNAGA *et al.*[1]	51	137	132
SCHLESINGER	89	197	178
SMITHIES and WALKER	18	39	197
STEINBERG *et al.*[1]	106	208	201
STEINBERG and MATSUMOTO	87	176	201 a
WALTER *et al.*	56	182	213
WENDT	306	693	216

[1] These authors give information on reciprocal matings.
[2] This reference includes the figures published separately by PROKOP and his colleagues [166, 167].

Table Va. Summary of results from the 16 family studies of the haptoglobin groups listed in Table V

Mating class	No. of families	No. of children Hp group				Total children
		1–1	2–1	2–2	0	
1–1 X 1–1	45	110	0	0	2	112
1–1 X 2–1	268	334	315	(1)	3	653
1–1 X 2–2	207	(3)	520	(6)	8	537
2–1 X 2–1	450	297	519	271	16	1103
2–1 X 2–2	779	(1)	947	878	34	1860
2–2 X 2–2	334	0	(2)	803	23	828
All groups	2083	745	2303	1959	86	5093
'Unknown' or						
0 X 1–1	12	10	25	0	1	36
X 2–1	38	21	56	28	0	105
X 2–2	33	0	13	30	3	46

Note: figures in parentheses are exceptions to simple Mendelian rules.

for the haptoglobin groups are surprising. There is a significant disturbance in the expected proportion of children in the mating class Hp 2–1 X Hp 2–2, but it is in the opposite direction to that previously found, Hp 2–2 children being present in excess.

Unfortunately, few of the published family studies have distinguished the phenotypes of mothers and fathers. Table Vb shows the results for the reciprocal mating classes in the four studies where such information is provided. It is interesting to note that although there is an excess of heterozygotes in the group 2–1 ♂ X 2–2 ♀, it is much more pronounced in the reciprocal group 2–2 ♂ X 2–1 ♀. The converse is true for the results obtained by MORTON *et al.* [135], the excess of heterozygotes being greater for the 2–2 ♂ X 2–1 ♀ matings.

Table Vb. Distribution of haptoglobin groups among children in reciprocal mating classes (based on the four studies marked[1] in Table IV)

Mating class Male	Female	No. of families	Total No. of children	Children's Hp group 1–1	2–1	2–2	0
1–1 X	2–1	24	83	40	42	(1)	0
2–1 X	1–1	15	39	19	20	0	0
1–1 X	2–2	28	85	0	82	0	3
2–2 X	1–1	29	81	(3)	72	(6)	0
2–1 X	2–2	108	255	0	131	117	7
2–2 X	2–1	88	256	(1)	143	107	5
Additional data from MORTON, KRIEGER and MI [135]							
2–1 X	2–2		610	–	288	322	–
2–2 X	2–1		605	–	273	332	–

Note: Figures in parentheses are exceptions to simple Mendelian rules.

Further analysis will be required to determine the validity of these mating-class effects, and particularly to attempt to explain the heterogeneity between the results based on European populations and those in Brazil.

Although the two-allele system satisfactorily explains the major part of the distribution of haptoglobin phenotypes among children shown in Table Va there is a number of exceptions. Specifically, one Hp 2–2 child to Hp 1–1 X Hp 2–1 parents, three Hp 1–1 and six Hp 2–2 children to Hp 2–2 X Hp 1–1 parents, one Hp 1–1 child to Hp 2–1 X Hp 2–2 parents and two Hp 2–1 children to Hp 2–2 X Hp 2–2 parents. These thirteen anomolous children all occur in two of the family surveys, eleven in that of HARRIS and his colleagues [91] and two in that of MATSUNAGA and his colleagues [132]. In the former study some of exceptions were attributed to illegitimacy, but for five of the children there was no evidence of this, and four of these in any case had haptoglobin types incompatible with that of their mothers: all five cases occurred in three closely related families. The two anomolous children in the Japanese study occurred also in one kindred, and there was also no evidence of illegitimacy.

Both HARRIS' group and MATSUNAGA and his colleagues believe that these 'atypical' segregations can be explained best by assuming that there is an inert allele, Hp^0, at the Hp locus such that Hp^1/Hp^0 persons are phenotypically Hp 1–1 and Hp^2/Hp^0 persons are phenotypically Hp 2–2 and Hp^0/Hp^0 persons are phenotypically Hp 0. The problem of the Hp 0 phenotype will be discussed in more detail below.

At the present time the hypothesis of an inert or 'silent' Hp^0 allele offers the most plausible explanation for most of these rare anomalous results in family studies of the haptoglobins. Two observations, however, seem relevant. The first is that atypical segregants have not been reported by other investigators. This suggests either that the Hp^0 gene has a very sporadic distribution, or that in other investigations atypical segregants have been excluded on the assumption of illegitimacy. This is stated in some of the studies, and the exclusions are supported by additional evidence based on blood grouping results [64, 67]. There still remains however a very substantial mass of family material in which the segregation of haptoglobin types in families is in accord with the hypothesis of the control of the common phenotypes by two codominant autosomal alleles Hp^1 and Hp^2. The exceptions only stress the need for further study to enligthen us on the mechanism controlling the atypical segregants.

b. Mother-Child Combinations

In many centres blood grouping and serum grouping is carried out routinely in cases of disputed paternity. The reliable information concerns the mother and her child, and it is possible to make limited use of the data collected in this manner. For the haptoglobin groups, for example, if the two-allele model is assumed to be correct we can predict that no Hp 1–1 mother can have an Hp 2–2 child and conversely no Hp 2–2 mother can have an Hp 1–1 child (in the previous section we have noted that on rare occasions exceptions may occur: in these cases extensive family studies may reveal the reason for the exception).

Mother-child combinations can be used also to test whether the distribution of the haptoglobin types among the children to mothers of any particular Hp group is in accord with expectation. If the Hp^1 and Hp^2 frequencies are known for the population to which the mothers belong the expected proportion of haptoglobin groups among their offspring may be calculated readily. Let p and q be the proportion of the autosomal alleles Hp^1 and Hp^2. Then the distribution of the phenotypes among children of mothers belonging to the three common haptoglobin groups will be:

Table VI. Expected proportion of children to mothers of various haptoglobin phenotypes. Frequency of Hp^1 = p and of Hp^2 = q

Mother's phenotype	Expected proportion of children 1–1	2–1	2–2
1–1	$p^2 + pq$	$pq + q^2$	0
2–1	$^1/_2\,p^2 + {}^1/_2\,pq$	$^1/_2\,p^2 + pq + {}^1/_2\,q^2$	$^1/_2\,pq + {}^1/_2\,q^2$
2–2	0	$p^2 + pq$	$pq + q^2$

Six published studies have included results for the haptoglobin groups of mother-child combinations. GALATIUS-JENSEN [69] collected 1035 cases, BECKMAN *et al.* [25] 564 cases, PROKOP *et al.* [168] 923 cases, FLEISCHER and LUNDEVALL [63] 294, MÄKELÄ, ERIKSSON and LEHTOVAARA [128] 222 and SCHLESINGER [178] a further 597 cases. GALATIUS-JENSEN found 148 of the children with an Hp 0 phenotype, approximately 14% of the total. Of the remainder, however, there were no mother-child exclusions and the distribution of the phenotypes among the children was very close to expectation. Similar results were obtained for the distribution of phenotypes by the five other laboratories; FLEISCHER and LUNDEVALL, SCHLESINGER and BECKMAN *et al.* do not report any Hp 0 children found in their surveys, but MÄKELÄ and his colleagues found only 1 in 222 and PROKOP's group 1 in 923 tested.

c. Twin Studies

If the haptoglobin groups are inherited in a simple Mendelian manner there should be no difference in the haptoglobin groups of monozygotic ('identical') twins, but dizygotic ('non-identical') twins may be different.

GALATIUS-JENSEN [69] and MÄKELÄ and his colleagues [128] have studied 496 pairs of twins with the following results.

Table VII

		No of pairs with Identical Hp-groups	Different Hp-groups	Total
Monozygotic	GALATIUS-JENSEN	126	0	
	MÄKELÄ *et al.*	61	0	187
Dizygotic	GALATIUS-JENSEN	50	30	
	MÄKELÄ *et al.*	133	69	282

Although 31% of the pairs of twins classified as dizygotic had different haptoglobin groups, none of 187 pairs classified as monozygotic differed in their haptoglobin groups. These results add little, of course, to our understanding of the formal genetics of the haptoglobin-group system, and are predictable on the basis of the genetic control of the haptoglobin groups substantiated by the family studies outlined above. On the other hand the argument may be used in a converse manner, for haptoglobin grouping is clearly a useful additional tool in helping to establish zygosity in twin studies being made for other purposes. The figures given above show that by using haptoglobin groups alone there is a 30% chance of confirming dizygosity—remembering that this applies of course only to populations where the haptoglobin gene frequencies are similar to that in northern Europe.

B. The Three-Allele Model

Two studies only have been made to test the genetic hypothesis put forward by CONNELL, DIXON and SMITHIES [50] that the Hp^1 gene can be resolved by sub-typing into two alleles Hp^{1F} and Hp^{1S}: these together with Hp^2 forming a three-allelic system. SMITHIES, CONNELL and DIXON [195] investigated only 21 families with 55 children and to this Goedde, Ritter and Weyrauch [81] have added a further 89 families with 153 children.

Since some of the parental combinations are relatively infrequent not all 21 possibilities are included among the 110 families summarized in Table VIII.

Table VIII. Distribution of children by haptoglobin sub-types in the combined studies of Smithies *et al.* [195] and Goedde *et al.* [81]

Parental phenotypes	No. of families	No. of children	Childrens phenotypes					
			1F–1F	1F–1S	1S–1S	2–1F	2–1S	2–2
1F–1F X 2–1F	2	2	1	–	–	1	–	–
1F–1F X 2–1S	4	7	–	5	–	2	–	–
1F–1F X 2–2	4	6	–	–	–	6	–	–
1S–1S X 2–1F	2	5	–	3	–	–	2	–
1S–1S X 2–1S	2	7	–	–	6	–	1	–
1S–1S X 2–2	1	2	–	–	–	–	2	–
1F–1S X 1F–1S	2	3	1	–	2	–	–	–
1F–1S X 1S–1S	1	3	–	1	2	–	–	–
1F–1S X 2–1F	6	9	1	4	–	3	1	–
1F–1S X 2–1S	5	10	–	4	–	–	6	–
1F–1S X 2–2	7	13	–	–	–	7	6	–
2–1F X 2–1F	6	9	–	–	–	7	–	2
2–1F X 2–1S	14	26	–	12	–	3	3	8
2–1F X 2–2	13	29	–	–	–	16	–	13
2–1S X 2–1S	6	9	–	–	3	–	3	3
2–1S X 2–2	22	44	–	–	–	–	24	20
2–2 X 2–2	13	24	–	–	–	–	–	24
Total	110	208	3	29	13	45	48	70

None of the 208 children tested are exceptions to the possible phenotypes predicted on the autosomal three-allele model, nor are there any marked deviations from expectation in the distribution of phenotypes among children for any of the parental mating classes. The total number of children in each class, however, is small and much more extensive studies will be required to examine the possibility of pre- or post-zygotic selection in the three-allele system.

C. Linkage

Few attempts have been made to determine close linkage between the genes at the haptoglobin locus and those for other serum protein, enzyme or blood group systems. HAUGE [92a] found no evidence for linkage between the Hp locus and the loci for the ABO, MNS, P, Rh, Lewis and Duffy blood groups or the gamma globulin groups. MÄKELÄ and his colleagues [128] in their study of 112 families state that there were several 'cross-overs' between genes at the Hp locus and those for the ABO, MN, Rh and Gm groups and also for those controlling ability to taste phenyl thio-carbamide (PTC), and they conclude that no linkage exists between these loci and that for Hp. Similarly, MATSUNAGA, MURAI and MATSUDA [132] using the sib-pair method of Penrose failed to detect any significant linkage between Hp and the ABO, MN, Rh and Lewis blood group systems.

A most stimulating suggestion on the possible location of the haptoglobin locus on one of the D-group chromosomes (No. 13, 14 and 15 of the Denver classification) was made by GERALD and his collaborators in 1964 and a more detailed report by the same group has been published more recently [71, 72]. They investigated a mentally retarded child whose haptoglobin group (Hp 1–1) was not consistent with that of its father (Hp 2–2), although careful study of the ABO, Rh, MNSs, Kell, Duffy, Kidd, Lutheran, Gm, Inv and Ag groups gave a very low probability to the possibility of non-paternity.

Cytological examination showed that the child possessed a 'ring' D Chromosome. Such chromosomes are thought to be produced by

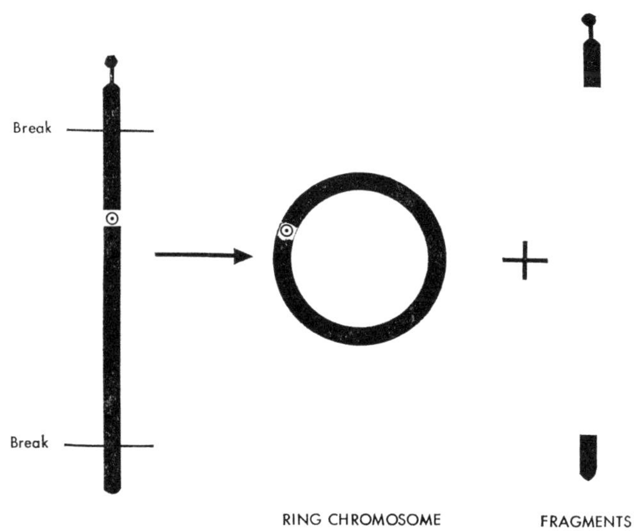

Fig. 6. Diagramatic representation of 'ring' chromosome formation.

deletion of terminal pieces of the chromosome followed by fusion of the two broken ends in the following manner (Fig. 6):

It is suggested that if the haptoglobin locus occurs near either end of this particular chromosome and the deletion of parts of the chromosome occurred during the formation of one of the father's sperms then the father would not have contributed his Hp^2 gene to the child. The Hp^1 gene in the child was contributed by her Hp 2–1 mother.

A second case of anomalous haptoglobin inheritance involving a 'ring' D chromosome has been reported by BLOOM and his co-workers [39]. The chromosome was identified as No. 13 and the interpretation of the haptoglobin finding suggested that the child had inherited a haptoglobin gene from his father but not from his mother. The authors conclude that this second case makes likely that the haptoglobin locus is near one end of chromosome No. 13. However, BIAS and MIGEON [35a] have studied a family with a short-arm deletion of the D_1 chromosome, which they believe is probably the one involved in the 'ring' D cases referred to above. Among the 14 persons carrying this short-arm deletion were some who were Hp 2–1. BIAS and MIGEON argue that this excludes the possibility of the Hp locus being on the short-arm of the D-chromosome deleted in their cases.

HUSTINX, BIJLSMA and NIJENHUIS [97] have investigated the possible location of the Hp genes near one end of a D chromosome by a different technique. In some individuals part of a D-chromosome becomes translocated to another member of the D group (D-D translocations), whilst in other individuals one of the No. 21 chromosomes becomes attached to one of the D group (D–21 translocations). The Dutch workers studied the distribution of the Hp groups in families where translocations involving the D chromosomes had been found and they concluded from the observed frequency of 'crossing over' of the haptoglobin genes that its location is unlikely to be on the long arm of the chromosome. Further studies of families in which deletions involving the D chromosomes have taken place could be valuable, but careful study to determine which D chromosome is involved in each case will be necessary to avoid confusion in interpretation of the results.

D. Forensic Applications of the Haptoglobin Groups

As with the various blood group systems, the haptoglobin groups have been examined for their potential usefulness in medico-legal cases.

The commonest application of such simple genetic systems is for excluding paternity in cases of disputed fatherhood. The principles of such applications have been dealt with in standard texts on human genetics [202], and will not be outlined here.

If testing is done for the three common haptoglobin phenotypes; Hp 1–1, 2–1 and 2–2 only, the possibility of exclusion is indicated below:

Table IXa. Possible exclusion of paternity using haptoglobin groups

Mother	Child	Father cannot be
1–1	1–1	2–2
	2–1	1–1
2–1	1–1	2–2
	2–1	–
	2–2	1–1
2–2	2–1	2–2
	2–2	1–1

Obviously if sub-typing is carried out the power of the method would be considerably increased (Table IXb).

For the Hp gene frequencies found in Denmark, GALATIUS-JENSEN [69] calculated the theoretical exclusion rate for paternity as approximately 18%. FLEISCHER and LUNDEVALL [63], however, in a study of 157 paternity cases in Oslo were able to exclude paternity by the use of haptoglobin typing alone in only 6, or 4% of the cases.

The usefulness of haptoglobin grouping in cases of disputed paternity has been questioned by both HARRIS *et al.* [90] and MATSUNAGA *et al.* [132]. They point out that the existence of atypical segregations which cannot be explained by illegitimacy must always raise doubts about the inheritance of the Hp groups in any particular family. Whilst this is true if reliance were to be placed solely on haptoglobin typing, an exclusion by Hp typing would certainly strengthen an exclusion already based on another system. The position may well be clarified in the future when the reasons for atypical segregation in rare cases is better understood.

Another potential forensic application is to aid in the identification of dried blood stains. CULLIFORD [56] has claimed that blood stains

Table IXb. Exclusions in paternity testing where sub-typing has been carried out

Mother	Child	Father cannot be
1F–1F	1F–1F	1S–1S, 2–2
	1F–1S	2–2
	2–1F	1S–1S
1F–1S	1F–1F	2–2
	1F–1S	2–2
	1S–1S	2–2
	2–1F	1F–1F, 1F–1S, 1S–1S
	2–1S	1F–1F, 1F–1S, 1S–1S
1S–1S	1S–1S	1F–1F, 2–2
	1F–1S	2–2
	2–1S	1F–1F
2–1F	1F–1F	1S–1S, 2–1S, 2–2
	1F–1S	1F–1F, 2–1F, 2–2
	2–1F	1S–1S
	2–1S	1F–1F, 2–1F, 2–2
	2–2	1F–1F, 1F–1S, 1S–1S
2–1S	1S–1S	1F–1F, 2–1F, 2–2
	1F–1S	1S–1S, 2–1S, 2–2
	2–1F	1S–1S, 2–1S, 2–2
	2–1S	1F–1F
	2–2	1F–1F, 1F–1S, 1S–1S
2–2	2–1F	1S–1S, 2–1S, 2–2
	2–1S	1F–1F, 2–1F, 2–2
	2–2	1F–1F, 1F–1S, 1S–1S

dried onto cloth can be treated by moistening with tris-citrate buffer and then inserting directly into the slot of a starch-gel. After electrophoresis and staining with benzidine reagent identification of the common haptoglobin patterns was possible.

4. THE PROBLEM OF Hp 2–1 (MOD) AND Hp 0

There is still no entirely satisfactory genetic hypothesis for the control of the two phenotypes Hp 2–1 (mod) or Hp 0. The 2–1 (mod) phenotype first described by CONNELL and SMITHIES [52] and by GIBLETT [73] has been reported as occurring sporadically in many parts of the world. In American Negroes, however, GIBLETT found nearly 10% of over 400 samples to possess this phenotype, though studies of black African populations show variable frequency of Hp 2–1 (mod). (See section 7 below.) Some of this variation in frequency, particularly the absence of Hp 2–1 (mod) in the early surveys of African populations, may have been due to initial failure to recognise this phenotype as a distinctive entity.

The pattern of the Hp 2–1 (mod) phenotype manifest after staining is also variable [GIBLETT, 74]. SUTTON and KARP [203] have proposed a subjective 5-point grading for Hp 2–1 phenotypes ranging from a normal 2–1 pattern, called 2–1a, to 2–1e, the latter showing a 1–1 band and only a faint trace band in the position of band 2 of the Hp 2–1 pattern. The four weaker grades, 2–1b to 2–1e, are considered by SUTTON and KARP to be 2–1 (mod) phenotypes, but they recognise also that there is no clear-cut separation between any of the grades.

Sub-typing of Hp 2–1 (mod) specimens reveals no abnormality of the hp 1 α-polypeptide chains, but there is a consistent decrease in the intensity of the Hp 2 α-polypeptide band. SMITHIES, CONNELL and DIXON [194] stated that the 2 α-polypeptide from 2–1 (mod) persons migrates between 1–2% slower than the normal hp 2 α-polypeptide. GIBLETT [74], however, considers that there is no difference in mobility, and that the variation resides solely in the quantity of the hp 2 α-polypeptide present in Hp 2–1 (mod) sera.

Hp 0, or a-haptoglobinaemia, does not in the majority of cases at least, represent a complete absence of haptoglobin. It should more correctly therefore be termed hypohaptoglobinaemia, except in those rare instances where even immunological methods fail to reveal the presence of haptoglobin. WHITTEN [217] was able to phenotype five out of six samples classified originally as Hp 0 after concentrating the serum by freeze-drying, and more recently MURRAY *et al.* [143] have used the technique of ultra-filtration to concentrate the haptoglobin to make possible the more definitive recognition of the Hp phenotypes. Unfortunately the routine application of such techniques to Hp 0 samples from American Negroes has not been reported. It is probable that if this were done many persons classified as Hp 0 could be assigned to another Hp phenotype.

Failure to detect haptoglobin is more common in young children than in adults although quantitative haptoglobin estimates show that near adult levels are reached by the age of one year [32]. Other factors, however, influence haptoglobin levels even in adults [41a]. NYMAN [153] reported originally that mean values were higher for Hp 1–1 than for Hp 2–1 which in turn was higher than Hp 2–2, and that in each case the mean value for males was higher than for females. Administration of oestrogens also affects Hp levels [42a]. Various diseases also influence the level of haptoglobin in the serum. (These are discussed more fully in section 6.) The changes due to disease, however, are reversible, and in normal healthy adults repeated testing over a period of several years suggests that serum haptoglobin levels are relatively constant, and the greater similarity in level between monozygotic compared to dizygotic twins [20] suggests that there are genetic factors involved in controlling the absolute quantity of haptoglobin in the serum.

A. Studies of American Negro Families

The occurrence of nearly 10% of Hp 2–1 (mod) and nearly 5% of Hp 0 phenotypes among American Negroes[1] prompted GIBLETT and STEINBERG [79] to investigate the patterns of inheritance of the haptoglobin groups in 92 Negro families. A further 91 Negro families has been investigated by SUTTON and KARP [203]. Both sets of data have been combined in Table Xa.

It should be noted, first of all, that there is a marked discrepancy between the proportion of Hp 0 among the children's phenotypes compared with the parents (13.8 and 3% respectively) though there is close agreement for the proportion of Hp 2–1 (mod) among the

[1] A later study in Georgia, USA found only 1.8% Hp 0 individuals among 167 Negroes tested [220].

Table Xa. Distribution of haptoglobin groups among children in negro families (GIBLETT and STEINBERG [79] and SUTTON and KARP [203])

Phenotypes of parents	No. of families	No. of children	Phenotypes of children 1–1	2–2	2–1	2–1 (Mod)	0
1–1 X 1–1	18	81	80	–	–	–	1
1–1 X 2–2	21	73	–	–	58	4	11
1–1 X 2–1	28	120	64	–	49	–	7
1–1 X 2–1 (Mod)	9	33	17	–	–	9	7
1–1 X 0	6	26	3	–	1	15	7
2–2 X 2–2	3	7	–	7	–	–	–
2–2 X 2–1	31	107	1	50	34	9	13
2–2 X 2–1 (Mod)	6	30	–	8	11	3	8
2–2 X 0	3	10	–	4	–	–	6
2–1 X 2–1	29	104	26	27	42	–	9
2–1 X 2–1 (Mod)	19	80	21	11	15	15	18
2–1 X 0	3	6	1	–	3	–	2
2–1 (Mod) X 2–1 (Mod)	5	20	3	1	–	10	6
2–1 (Mod) X 0	2	7	1	1	3	–	2
0 X 0	0	0	–	–	–	–	–
	183	704	217	109	216	65	97

children and parents (9.2 and 10.7%). This increase in frequency of the Hp 0 phenotype in children has been found also in many of the family studies involving non-Negro families shown in Table V, though the overall frequency (1.8%) is lower for non-Negro children. This may merely reflect a lower mean haptoglobin concentration in the sera of Negroes than in non-Negro sera, but no relevant figures have been published. Secondly, it should be noted that the Hp 1–1 child from an Hp 2–2 X Hp 2–1 mating in Table Xa is considered by SUTTON and KARP to be illegitimate.

GIBLETT and STEINBERG believe that the results of their Negro family study can best be interpreted by assuming another allele, Hp^{2m} at the haptoglobin locus such that the relationship between genotypes is as follows:

Genotype		Phenotype
Hp^{2m}/Hp^{2m}	results in	Hp 2–2 or occasionally Hp 0
Hp^{2m}/Hp^{2}	results in	Hp 2–2 or occasionally Hp 0
Hp^{2m}/Hp^{1}	results in	Hp 2–1 (mod) or Hp 0

Certain predictions can be made about the occurrence of Hp 2–1 (mod) children if these assumptions are correct.

1. Matings Hp 1–1 X 1–1 or 1–1 X 2–1 will not give Hp 2–2 or 2–1 (mod).
2. Matings Hp 2–2 X 2–2 will not give 1–1 or 2–1 (mod).
3. Matings Hp 2–1 X 2–1 will not give 2–1 (mod).
4. Matings 1–1 X 2–1 (mod) will not give 2–2 or 2–1.
5. All other mating classes may produce some 2–1 (mod) offspring.

The figures in Table Xa support these predictions, and SUTTON and KARP believe that the GIBLETT and STEINBERG hypothesis is essentially correct as an explanation of the control of the Hp 2–1 (mod) phenotype. They believe also, however, that the Hp 2–1 (mod) phenotype can be divided further, and they suggest that there are two alleles Hp^{cd} and Hp^{c} which depress the formation of the hp 2 polypeptide, Hp^{c} having the larger effect.

Another approach to the problem of the occurrence of the Hp 2–1 (mod) and Hp 0 phenotypes in Negro families was made by PARKER and BEARN [160, 161]. Stimulated by the model proposed by JACOB and MONOD [100], for the control of the β-galactosidase enzyme system in *Escherichia coli* they proposed that there were two closely linked loci involved in the control of haptoglobin synthesis, one locus representing a control gene, *C*. and the other the structural gene *Hp*. The control gene produces an activator, A, which diffuses to the site of the structural gene and switches it on so that haptoglobin synthesis can be initiated. In PARKER and BEARN's hypothesis there are two controller genes, C^1 and C^2 producing activators A_1 and A_2 which in turn switch on the Hp^1 or Hp^2 genes. Their hypothesis also assumes that A_2 can act equally to switch on Hp^1 and Hp^2, but A_1 is much more active in switching on Hp^1 than Hp^2.

It follows that a mutation in C^1, (C^{1-}), will be ineffective if C^2 is normal. The latter gene will continue to produce A_2, which in turn will switch on both Hp^1 and Hp^2 equally. In such a heterozygote a normal Hp 2–1 pattern will result. On the other hand, a mutation of C^2 to give C^{2-} will prevent the formation of A_2, and in the presence of a normal C^1 gene A_1 will switch on Hp^1 but will be less active in switching on Hp^2. There will therefore be less Hp^2-gene product to combine with Hp^1-gene product, with a resulting excess of Hp 1 and a decrease of the Hp 1–2 polymers. Thus the pattern resulting from the heterozygous combination $C^1Hp^1/C^{2-}Hp^2$ will be Hp 2–1 (mod). Finally, the combination of the mutant C^{1-} and C^{2-} will lead to complete failure of production of haptoglobin, giving the Hp 0 phenotype. Figure 7 will help to make these relationships clear.

The assumptions made by PARKER and BEARN explain the depression of the Hp^2 gene product, which has been demonstrated subsequently as a decreased production of the Hp 2 α-polypeptide chain. However, when the model is tested against the pedigrees of GIBLETT and STEINBERG's Negro families not only is it necessary to assume

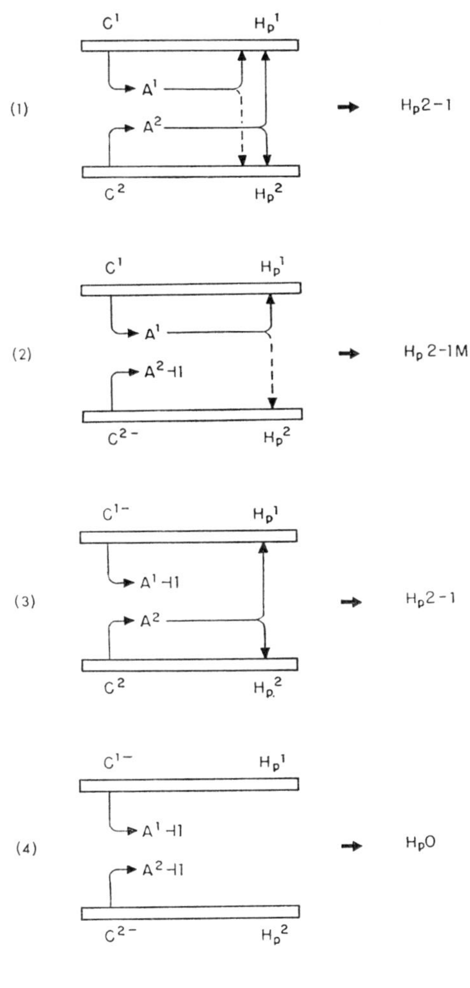

Fig. 7. The effect of controller-gene mutations in the production of haptoglobin according to the hypothesis of PARKER and BEARN [161].

that there were more cases of illegitimacy than for the hypothesis of a mutation to the Hp^{2m} allele, but in order to get approximate gene frequency estimates agreeing with the observed distributions it has to be further assumed that the C^{2-} mutation is much more frequent than the C^{1-} mutation.

The control of the Hp 0 phenotype in Negro families is more difficult to explain. As noted above, the frequency of Hp 0 among Negro children is nearly 12% higher than among their parents, and it has to be assumed that these 12% will at least have recognisable Hp phenotypes as adults. GIBLETT and STEINBERG postulated that Hp 0 was nearly always an expression of the genotype Hp^{2m}/Hp^1, but that occassionally it was Hp^{2m}/Hp^{2m} or Hp^{2m}/Hp^2. These assumptions fit the data in Table Xa, remembering that sometimes, even in non-Negro families, children of any genotype may appear to be Hp 0, though this is uncommon in the mating class Hp 1–1 X Hp 1–1.

On PARKER and BEARN's hypothesis the Hp 0 phenotype in Negroes will result from the inheritance of the two controller gene mutations C^{1-} and C^{2-}. Such a person of course will be invariably Hp 0. At present it is not possible to discriminate among Hp 0 children in Negro families those which later will develop a recognisable Hp pattern. Longitudinal studies could help in this respect.

The attempts to find a rigorous genetical explanation of both the Hp 2–1 (mod) and Hp 0 phenotypes in Negroes may be misguided. The increased frequency of Hp 0, particularly among Negro children, may merely represent the larger effect of the Hp^{2m} compared to the Hp^2 allele in depressing haptoglobin levels more frequently below the point where recognition of the Hp pattern becomes possible. If this is so we may expect to find the percentage of Hp 0 children in any mating class to be proportional to the number of Hp^2 or Hp^{2m} alleles present among pairs of parents in the following order of increasing frequency:

4 Hp^1, 3 Hp^1 1 Hp^2, 2 Hp^1 2 Hp^2, 1 Hp^1 3 Hp^2, 4 Hp^2, and that for each Hp category an Hp^{2m} allele replacing an Hp^2 allele will result in an increased frequency of Hp 0 among the children.

From Table Xb, which is based on the combined results of the Negro families studied by GIBLETT and STEINBERG and SUTTON and KARP, it can be seen that the addition of an Hp^2 allele among parents increases the frequency of Hp 0 children from 1% to nearly 6%, whilst the replacement of Hp^1 with two or three Hp^2 increases the frequency to 11 to 12%. Replacement of any one of the Hp^2 alleles by an Hp^{2m}

allele increases the frequency considerably, and with two Hp^{2m} alleles present among the parents, 30% of the offspring are Hp 0.

These results suggest most strongly that the greater proportion of Hp 0 phenotypes among the children of Negro families is related to a simple dosage effect depending on the number of Hp^2 and Hp^{2m} alleles possessed by the parents. The only discrepant result in Table X b is that where both parents are Hp 2–2, *i.e.* possessing 4 Hp^2 alleles. The total number of children, however, in this mating class is only 7, and the number of Hp 0 children expected is only two. The discrepancy therefore may be due to sampling error.

Table Xb. Number and percentage of Hp 0 children to parents with varying numbers of Hp^1, Hp^2 and Hp^{2m} alleles. Based on Negro families in Table X a

No. of genes among parents	Total No. of children	No. of Hp 0 children	Percentage of Hp 0 children
4 Hp^1	81	1	1.2
3 Hp^1 : 1 Hp^2	120	7	5.8
2 Hp^1 : 2 Hp^2	177	20	11.2
1 Hp^1 : 3 Hp^2	107	13	12.1
4 Hp^2	7	0	0
3 Hp^1 : 1 Hp^{2m}	33	7	21.2
2 Hp^1 : {1 Hp^2, 1 Hp^{2m}}	80	18	22.5
1 Hp^1 : {2 Hp^2, 1 Hp^{2m}}	30	8	26.6
2 Hp^1 : 2 Hp^{2m}	20	6	30.0

B. Hp 0 in Non-Negro Families

Some individuals in all populations, as has been noted above, have insufficient haptoglobin present to enable the Hp phenotype to be assigned by usual typing on starch-gel. In some populations many of these persons, as for instance in parts of S.E. Asia or New Guinea [57, 114], probably have low haptoglobin levels because of endemic malaria or other diseases resulting in haemolytic episodes. There remains, however, a small number of persons in apparent good health whose low haptoglobin level may be genetically determined. In some of these cases the collection of family pedigrees shows that other members of the family may also have low haptoglobin levels and be recorded as phenotypically Hp 0, (GOTTLIEB *et al.* [85, 86]). This familial concentration of Hp 0 persons lends support to the possibility of a simple genetic control.

In addition to families showing more than one person with the Hp 0 phenotype, there are others which show a segregation of phenotypes which is not in accord with the simple two-allele hypothesis for the inheritance of the haptoglobin groups. In the case of atypical segregations reported by HARRIS *et al.* [90, 91] five examples occurred in three closely related families. Four of the exceptional offspring were incompatible with their mothers' haptoglobin types, which rules out the possibility of illegitimacy as an explanation of the discrepancies. The kindred in which these five atypical segregants occurred contained also an Hp 0 individual. HARRIS and his colleagues believed this individual might be homozygous for a rare allele Hp^0 which suppresses haptoglobin production more or less completely.

On this hypothesis persons of genotype Hp^1/Hp^0 will appear phenotypically as Hp 1–1, and of genotype Hp^2/Hp^0 will appear phenotypically as Hp 2–2. The appearance of atypical segregants within the kindred can then be readily explained since, for example the mating between an Hp 2–1 and Hp 1–1 person which gave an Hp 2–2 child could be genotypically

$$Hp^2/Hp^1 \quad x \quad Hp^1/Hp^0$$
$$Hp^2/Hp^0$$

MATSUNAGA [130] provided another kindred with atypical segregation in two generations, and in which at least one person showed low haptoglobin levels. His atypical segregants could be explained also on the basis of the existence of a 'silent' *Hp* 0 allele, and he calculated its probable frequency in the Japanese population and estimated that one person in 2500 would be Hp 0 because they were homozygous for this allele.

PARKER and BEARN [161] in discussing their control gene hypothesis realised that the mechanism suggested for the genetic control of Hp 2–1 (mod) and Hp 0 was not satisfactory for explaining the occurrence of Hp 0 in populations where Hp 2–1 (mod) was not present in the same kindreds. They suggested that in such populations there was a single controller allele C and a mutation C^- which inhibited production of the activator A. Homozygotes for the C^- mutation would be Hp 0, but heterozygotes C^-/C would have normal phenotype. They realised it is not possible to distinguish between their model and that of a recessive suppressor gene as postulated by SUTTON *et al.* [205].

More recently MURRAY, ROBINSON and VISNICH [143] have studied several families in which either a parent or a child had been found to have low haptoglobin levels. In five families in which both parents had normal haptoglobin levels, 7 out of 13 offspring had low levels. In two other families one parent and one child had low levels, and in the remaining three families a parent only was Hp 0. These patterns are similar to those to be expected for a trait controlled by a recessive gene which is not too uncommon in the population, but at the same time they can also be interpreted as an example of dominant inheritance with less than 100% penetrance.

Of greater interest, however, is one family in which the mother is Hp 0 and the father Hp 2–2, with one child Hp 2–2 and the other Hp 2–1. If both these children are legitimate the Hp 0 parent must have contributed an Hp^1 allele to one child and an Hp^2 allele to the other, that is she must herself be genetically Hp^1/Hp^2. Such a result rules out the possibility, at least in this instance, of an Hp^0 allele. It is compatible with PARKER and BEARN's recessive controller gene hypothesis, the mother in this case being $C\text{-}Hp^1/C\text{-}Hp^2$ and the father CHp^2/CHp^2.

Another recent paper by HIGASHI and LUBS [96] reports an extensive kindred containing several persons with low or absent haptoglobin in the serum. There were no atypical segregants in the pedigrees, but there is one phenotypically Hp 0 child from an Hp 2–1 X 2–1 mating which again cannot be explained on the basis of a 'silent' Hp^0 allele. One of the other persons with a low haptoglobin level manifested a pattern similar to SUTTON and KARP's 2–1d pattern. HIGASHI and LUBS concluded that a significant proportion of the quantitative and phenotypic variations described by them is due to the interaction of environmental and age effects with various Hp^{2m} alleles of the type suggested by SUTTON and KARP.

Clearly, further family studies, including quantitative and longitudinal investigations are needed before a complete understanding of the factors controlling the appearance of Hp 0 individuals is obtained.

5. THE STRUCTURE AND FUNCTION OF HAPTOGLOBIN

A. Physico-Chemical Studies

Interest in the haptoglobins dates back to 1938 when POLONOWSKI and JAYLE [164] noticed that the peroxidase activity of haemoglobin was enhanced by the addition of serum. This observation led to the discovery that haemoglobin was bound irreversibly with a component of serum which was later isolated by JAYLE and his colleagues, and which they called haptoglobin [89, 103–105]. They found that both by free electrophoresis or paper electrophoresis haptoglobin migrates with the α_2 fraction of plasma, which constitutes about 10% of the total plasma proteins. The actual amount of haptoglobin varies from person to person, but on average is roughly about 100 mg per 100 ml of serum and constitutes about 20% of the α_2 fraction.

JAYLE's group showed that two forms of haptoglobin could be obtained by purification from different persons: the original pure haptoglobins were obtained from the urine of nephrotic patients. Both forms were glyco-proteins with about 83% protein, one having a molecular weight of about 85,000 (HERMAN-BOUSSIER *et al.* [93] have revised this figure to 100,000, whilst estimates made from the Hp-Hb complex suggest 89,000) and the other a molecular weight approximately 170,000. It was concluded that the one was a monomer, the other a dimer, and they showed also that the former combined with haemoglobin in equimolecular proportions whilst the latter binds two molecules of haemoglobin to each molecule of haptoglobin.

Purification of haptoglobin was also carried out by LAURELL [121], using as a source ascitic fluid. Some of the difficulties of preparing pure haptoglobin by salting-out procedures are referred to by LAURELL and GRONVALL [123]. Subsequently, a method of preparing pure haptoglobin from normal serum using column chromotography has been worked out in detail by CONNELL and his colleagues [51], and this has formed the basis for most recent studies of haptoglobin structure.

The identification by SMITHIES [188, 189] of the three distinctive haptoglobin group patterns, Hp 1–1, 2–1 and 2–2 led to a re-examination of the purified haptoglobins obtained from the α_2 plasma fraction. BEARN and FRANKLIN [22] showed that not only could one discriminate the haemoglobin-binding protein from Hp 1–1, 2–1 and 2–2 sera by electrophoresis in starch block, but also the pattern of each was distinctive in the ultracentrifuge. This was important, because it implied that the heterozygote Hp 2–1 individual was producing a distinctive molecule and not just a mixture of 1–1 and 2–2 molecules.

The results referred to above led ALLISON [3] and also SMITHIES [190] to deduce that the multiple bands which are observed after electrophoresis in starch-gel of sera from Hp 2–2 and Hp 2–1 persons could result from a tendency of the Hp^2 gene product to polymerize and also to combine with the Hp^1 gene product, where this was available. ALLISON's schema is shown below (Fig. 8):

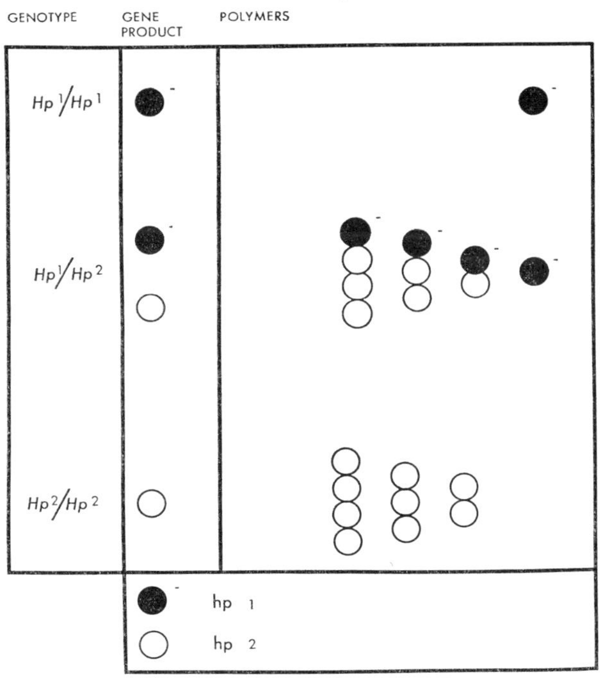

Fig. 8. The production of haptoglobin polymers according to the scheme of ALLISON [3].

It was suggested that the Hp^1 gene product, of molecular weight about 80,000, has only a single combining site and does not form polymers. It is also negatively charged and on electrophoresis in starch-gel it migrates toward the anode as a single band. The gene product of Hp^2 on the other hand has complementary binding sites and also a lower negative charge than the product of Hp^1. In Hp^2/Hp^2 individuals, therefore, the proteins produced can combine with one another to form polymers; a small proportion of dimers, a slightly higher proportion of trimers, and decreasing proportions of tetramers and higher polymers resulting.

In heterozygotes Hp^1/Hp^2 both hp 1 and hp 2 molecules are produced and these form mixed polymers, each containing only one hp 1 molecule. If hp 1 has a greater negative charge relative to hp 2 these polymers will migrate faster than the corresponding hp 2 polymers. Further, because of the polymerization of the hp 2 molecule and the combination of these polymers with only one molecule of hp 1 in each case there will be an excess of hp 1 molecules which will migrate as a single band in the Hp 1–1 position. The polymers present in Hp 2–1 and Hp 2–2 sera are apparently stable, for isolation of individual bands followed by further electrophoresis in starch gel results in single bands occupying the position of the one isolated. In addition JAVID [101] has shown that the slower moving components of Hp 2–1 after sub-typing reveal the presence of the hp 1 polypeptide, whereas this is not true for the slower moving components from Hp 2–2.

More recent studies have focussed attention on the breakdown of the purified haptoglobin into its constituent polypeptide chains. The original work of SMITHIES and his colleagues referred to in Section 1 above has been followed by more precise definition of the conditions for separating the α and β polypetide chains [SMITHIES *et al.*, 196; CONNELL *et al.*, 53].

The α and β-polypeptides are obtained by treating the purified haptoglobin with 2-mercapto-ethanol in the presence of 8m urea to bring about cleavage of the disulphide bonds holding the chains together, followed by alkylation with 0.05 m iodo-acetamide. The α-chains may then be separated from the β-chains by differential precipitation with sodium dodecylsulphate. Ultra-centrifuge analysis gives a molecular weight of 8,900 ± 400 for the hp 1 α-chains and 17,300 ± 1,400 for the hp 2 α-chain. CHEFTEL and MORETTI estimate the molecular weight of the hp β polypeptide as 40,000 [47b] In native haptoglobin each hp 1F α or hp 1S α-chain contains 1 disulphide bridge, whereas each hp 2 α-chain contains two such bridges. This observation fits in with the hypothesis that the hp 2 α-polypeptide results from an approximate duplication of the *Hp*[1] cistron.

No difference can be observed in the electrophoretic behaviour or peptide structure of the hp β-chains from any of the three common haptoglobin phenotypes [49a]. Starch gel electrophoresis reveals six bands of lower mobility than either the hp 1 α or hp 2 α-bands. Isolation and chemical characterization of the β-polypeptide has not so far been as successful as for the α-polypeptides.

Treatment of the isolated chains with neuraminidase to remove the sialic acid residues results, however, in an increase in the mobility of the β-chain, but not of the α-chain. On the basis of these experiments Shim and Bearn [185] conclude that sialic acid is bound exclusively to the β-chain. Direct analysis for sialic acid by the modified orcinol method confirmed the conclusion, for although the β-chain contained approximately 7.5% of sialic acid, the purified α-chain preparations were found to be free.

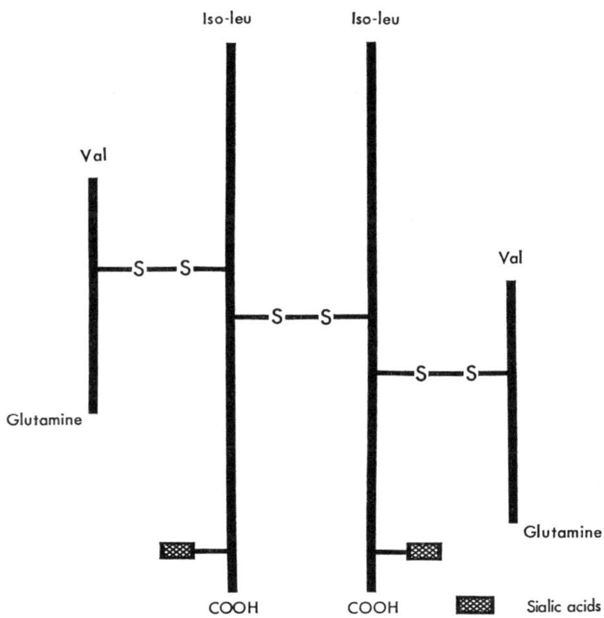

Fig. 9. Shim and Bearn's [185] model of the hp 1 molecule. The α-chains are shown on the outside and the β-chains on the inside of the model. Sialic acid residues are attached to the β-chains.

Shim and Bearn [185] found also that the isolated α-chain comprised approximately 19% of the hp 1 molecule. Since the molecular weight of the hp 1 polypeptide is almost 9,000, and if we accept a value of 100,000 for the molecular weight of the intact hp 1 molecule, it suggests that each molecule has two α-chains. Smithies, Connell and Dixon [194] found the N-terminal amino-acid of the α-chain to be valine, and Smith, Edmon and Owen [187] have subsequently shown that there are equi-molecular proportions of valine and isoleucine as the N-terminal amino-acids of the intact molecule, suggesting further, that there are two β-chains as well.

Shim and Bearn [185] point out also that although there are perhaps 16–18 half-cystine residues in the intact molecule, the work of Robert, Boussier and Jayle [174] indicates that about nine sulfhydryl groups are released in the presence of guanidine hydrochloride, or that there are perhaps three or four disulphide bonds in the intact molecule. The combined evidence on molecular weights, N-terminal amino acid and sialic acid analysis and estimation of sulfhydryl residues led them to propose the following tentative structure for the hp 1 molecule (Fig. 9).

B. Immunological Reactions of Haptoglobin

Another approach to an understanding of differences in molecular structure is the use of specific antisera to detect differences in the antigenic structure of protein molecules. Attention was first drawn to the antigenic properties of haptoglobin by Burtin *et al.* [46], and after the discovery of the three common haptoglobin phenotypes both Fine and Battistini [62] and Bearn and Franklin [23] prepared antisera against haptoglobin by injecting rabbits with whole human serum and separated haptoglobin-haemoglobin complexes respectively. Neither group, nor that of Schulze *et al.* [180], was able to detect any difference in the precipitin reactions, using agar-gel diffusion techniques, between the anti-serum and Hp 1–1, 2–1 or 2–2 sera. A similar result was obtained with an antiserum prepared by Aly and his colleagues in Germany [9, 10], with the exception of the reaction shown by serum from one person, a patient with cirrhosis of the liver. Serum from this individual tested against the rabbit anti-haptoglobin serum gave, in addition to a precipitin arc indicating identity of reaction with normal Hp 1–1, 2–1 or 2–2 sera, a small spur on the arc indicating the presence of antigenic components in the patient's serum not present in normal sera. Similar spurs were found in a brother and a son of the propositus, both of whom were normally healthy.

When the anti-haptoglobin serum was absorbed with normal Hp 2–1 serum the precipitin reaction with Hp 1–1, 2–1 and 2–2 sera was abolished, but it still reacted with sera from the three atypical individuals detected originally. It was concluded that these three persons had each inherited a rare gene which altered the antigenic determinants of the haptoglobin molecule. It was pointed out, however, that these same determinants must have been present to some extent in the original sera used for producing the antiserum, since this was capable of reacting with them. The authors concluded therefore that their atypical cases may merely represent individuals with a larger than normal dose of these determinants, and that these are more commonly present in the population though in a weaker form.

SHIM and BEARN [185], using an antiserum prepared against purified type 1–1 haptoglobin also found several instances of atypical immunological reactions in a panel of Nigerian sera, and these reactions were independent of the starch-gel electrophoretic Hp phenotype. Similar atypical immunological reactions were noted also by KORNGOLD [117]. KORNGOLD showed further, that if antisera were prepared in rabbits against purified haptoglobin from Hp 2–2 sera they could differentiate between Hp 2–2 and 2–1 on the one hand and Hp 1–1 on the other: the greater the degree of purification the better the discrimination. The lack of differentiation previously, therefore, was due apparently to lack of purity of the antigen used for immunization, and consequent weaker reactions.

Anti-Hp 2–2 serum when absorbed with an Hp 1–1 serum lost its ability to react in the double diffusion test with Hp 1–1 sera, but still reacted, though more weakly, with Hp 2–2 and 2–1 sera. When the absorbed antiserum was tested also against the bands resulting after electrophoresis in starch gel of an Hp 2–1 serum, there was no reaction against the fastest band (presumably containing only hp 1 molecules) but there was a reaction against the slower bands. This latter reaction however was less intense than that obtained against the corresponding bands of an Hp 2–2 serum.

KORNGOLD's work has been extended further by CLEVE *et al.* in Germany [BEURING *et al.*, 34, 35 and EICHMANN *et al.*, 60]. Using antisera prepared against both Hp 1–1 sera and purified haptoglobin from Hp 2–2 sera and tested against both native haptoglobins in

Hp 1–1, 2–1 and 2–2 sera and against the haptoglobins after formation of the Hp-Hb complexes they concluded that there are at least three recognizable antigenic components associated with the haptoglobin molecule.

Antigenic determinant A is present in all haptoglobin molecules, and it can be detected both in free haptoglobin and in Hp-Hb complexes from all three groups Hp 1–1, 2–1 and 2–2.

Antigenic determinant B can be detected also in free haptoglobin from all three groups, but it cannot be detected in Hp 1–1 sera after complexing with haemoglobin. The reaction with antigen B still remains in complexed Hp 2–1 and 2–2 sera, but it is weaker than with the free haptoglobins of these groups. This would suggest that the active site for the B determinant is covered by haemoglobin when Hb is bound to the hp 1 molecule, but that the surface configuration of the hp 2 molecule is such that the B-site is left partially uncovered when the Hp-Hb complex is formed.

The C antigenic determinant is characteristic of the hp 2 molecule. Reactions against C are present in Hp 2–2 and 2–1 sera but absent in Hp 1–1 sera. The reactions are partly inhibited by binding with haemoglobin.

The antigenic characteristics of the α and β-polypeptide chains have been examined also by SHIM and BEARN [185] and by SHIM and his colleagues in Korea [186]. Using an antiserum prepared in rabbits against a purified hp IF and hp IS mixture antigenic determinants could be detected on both the hp 1 α and β-chains. Similarly antisera prepared against purified haptoglobin from Hp 2–2 sera were tested also against the hp 2 α and β-chains, and antisera against free haptoglobin and Hp-Hb complexes.

Their results with anti-hp 2 serum were similar to those of KORNGOLD and of CLEVE *et al.* outlined above, confirming that the hp 2 molecule has antigenic determinants not present in the hp 1 molecule. Their anti-hp 2 serum reacted also with both free and complexed haptoglobin in Hp 2–2 sera. However, anti-β-chain serum reacted only with free haptoglobin, and not with the Hp-Hb complex, suggesting that haemoglobin is bound to the β-chain. Finally, anti-β-chain serum reacted equally well with free haptoglobin in sera from Hp 1–1, 2–1 and 2–2 persons. This suggests that the difference noted between Hp 1–1 and Hp 2–2 sera by KORNGOLD, or the C antigen of the German workers, is part of the structure of the α-polypeptide.

All of this intriguing work on the antigenic structure of the haptoglobins agrees well with what may be surmized from the more direct chemical studies. No differences have been demonstrated so far in the β-chains from the three common haptoglobin phenotypes, either by starch gel electrophoresis of the separated chains or by amino acid analysis or fingerprinting of the purified chains [CLEVE, 49a]. The hp 2 α-chain however is not only twice the size of the hp 1F or hp 1S α-chains, but it contains a peptide (j) not present in the tryptic digests of either the 1F or 1S α-chain. This could readily be responsible for the C antigen reaction of the hp 2 molecule.

The location of the A and B antigens is not clear. If haemoglobin is bound to the β-chain the fact that the reaction with A is not affected by binding with haemoglobin suggests that it is on the α-chain, but in a position distant from the 'J' peptide. The B antigenic determinant, on the other hand, is on the β-chain and is masked by binding with haemoglobin. It is not clear why the C antigen, if it is part of the hp 2 α-chain, is also partly masked in the Hp-Hb complex. Further study will be needed to clarify this problem.

SHIM *et al.* [186] have deduced from the relative percentage of the hp 2 α-chains present in hp 2 α molecules that there are equal numbers of hp 2 α and hp β-chains present in hp 2 molecules. They postulate that if the molecular weight of the hp 2 α-chain is 18,000 and that of the hp β-chain is 36,000 the bands of the Hp 2–2 type represent the following molecular associations:

$$(2\ \alpha{-}\beta)_4 : (2\ \alpha{-}\beta)_6 : (2\ \alpha{-}\beta)_8 : (2\ \alpha{-}\beta)_{10}$$

The hp 1 molecule on the other hand they regard as $(1\ \alpha{-}\beta)_2$. In the heterozygote the bands would be:

$$(1\ \alpha{-}\beta)_2 : (1\ \alpha{-}\beta)_2\ (2\ \alpha{-}\beta)_2 : (1\ \alpha{-}\beta)_2\ (2\ \alpha{-}\beta)_4$$
$$: (1\ \alpha{-}\beta)_2\ (2\ \alpha{-}\beta)_6$$

These associations represent the polymers in descending order of mobility in starch gel.

C. The Haptoglobin-Haemoglobin Complex

Finally, brief mention must be made of the binding of haptoglobin with haemoglobin. It was pointed out above that the immunological evidence suggests that Hb binds with the β-chain of haptoglobin. GORDON and BEARN have demonstrated also that haemoglobin binds with isolated β-chains but not with α-chains [83a]. NAGEL and RANNEY [146] have shown that haptoglobin binds with Hb A; Hb F; Hb A_2; Hb I and Hb Lepore, all of which contain α-chains, but not with Hb H and Hb Barts, which do not contain α-chains. However, GIBLETT [75] has recently reported that haptoglobin does not bind with purified haemoglobin α-chains. It seems likely therefore that the two unlike haemoglobin chains, α and β, are both necessary for complex formation with haptoglobin to occur.

Since hp 1 molecules bind equimolecular proportions of haemoglobin SHIM and his colleagues [186] believe that the actual bond is made with a half molecule of haemoglobin. This suggestion had been made previously by LAURELL [122] who postulated that the force binding the haemoglobin to haptoglobin was stronger at neutral pH than the force binding the two halves of the haemoglobin molecule together. In the case of the hp 1 molecule this could be represented diagrammatically, as in Figure 10.

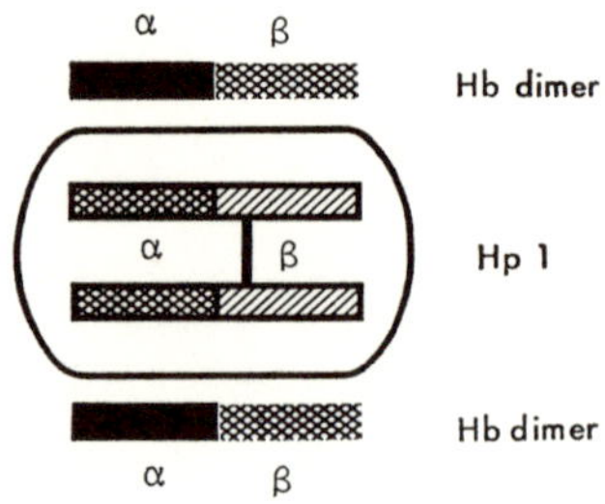

Fig. 10. Diagram of the association between the haemoglobin dimers and the hp 1 molecule as postulated by SHIM *et al.* [186].

If this is true, the intermediate complex which is formed with haemoglobin would represent the binding of only half a molecule of haemoglobin per molecule of haptoglobin. JAYLE *et al.* [108] prefer to think of this intermediate complex as one molecule of haemoglobin binding to two molecules of haptoglobin.

Using the method of WINTERHALTER and HUEHNS [219], Giblett's laboratory [75] has prepared an α β globin preparation from haemoglobin A and added it to purified hp 1. A single Hp-globin complex was formed migrating in starch-gel at the same rate as the saturated Hp-Hb complex. However subsequent addition of haem to a mixture of globin and excess haptoglobin resulted in the formation of an intermediate Hp-Hb complex as well as the saturated complex. When globin was in excess only the saturated complex was formed. GIBLETT concludes that not only do these results demonstrate that the hp molecule can bind to the α β haemoglobin dimer but that the addition of haem is necessary to allow the formation of the intermediate complex. This may be due to alteration of the dissociation constant between hp and globin after the addition of haem, requiring a higher concentration of Hb to saturate both binding sites of the hp molecule.

6. THE FORMATION AND FUNCTION OF HAPTOGLOBIN

Little is known about the site of formation of haptoglobin. The studies of Nyman and of others, [Nyman, 153; Owen *et al.*, 156] show that patients with severe liver disease are deficient in haptoglobin, which suggests that the liver may be the site of formation. However since such patients have lowered red-cell survival times, the depletion of haptoglobin may be due to intravascular red-cell destruction. Whatever its source it starts to be produced early in foetal life and Giblett has reported a case of twins born prematurely with nearly normal haptoglobin levels. Galatius-Jensen [69] suggested that the low concentration of serum haptoglobin found in newborn infants may be due to haemolysis of foetal red-cells rather than inability to synthesize the protein. The levels early in infancy are certainly lower than in adults, rising to maximum value by about the 20th year. Nyman [153] points out that depletion of haptoglobin is not followed by any rapid increase in haptoglobin production. On the other hand, tissue damage or infection is a very potent stimulant.

Although it is difficult to measure the half-life of haptoglobin, indirect estimates based both on the increase to normal levels following injection of haemoglobin to bind existing haptoglobin and the fall to normal levels following the rise after infection or injection of a pyrogen suggests a half-life of between 3 to 7 days [Laurell and Nyman, 124]. Murray and Connell [140] artificially increased haptoglobin levels in rabbits by injecting turpentine and found a halflife of approximately $2^1/_2$ days.

Ideas on the possible function of haptoglobin have been centred on its binding capacity for haemoglobin [154]. The demonstration of this irreversible binding in serum was followed by direct observations on humans by Laurell and Nyman [124]. Injection of haemoglobin was followed by rapid depletion of the haptoglobin in the blood, with the formation of a haptoglobin-haemoglobin complex. This complex does not appear in the urine of healthy persons, although small quantities of free haptoglobin of type 1–1 and the smaller molecules of type

2–1 and 2–2 can be detected after concentration in normal urine [BERGGAARD and BEARN, 31]. The Hp-Hb complex is presumably removed from the circulation by the reticulo-endothelial system and in the rabbit, MURRAY, CONNELL and PERT [141] have found that most of the complex was localized in the liver. KEENE and JANDL [111] have shown that in man haemoglobin labelled with Fe^{59} is removed mainly by the liver, moderately by the bone marrow and relatively little by the spleen and other organs. Binding by haptoglobin does not effect this distribution.

The rate of removal of the complex in human subjects was found by LAURELL and NYMAN [124] to be linear and had a maximum value of about 13 mg per 100 ml per hour of haptoglobin. LATHEM [118] has found some subjects in which the rate of removal of the complex is exponential, although in others linear rates were observed.

Almost simultaneously LAURELL and NYMAN [124], and ALLISON and AP REES [4], suggested that haptoglobin determines the renal threshold for haemoglobin, its excretion occurring only after all the haptoglobin in the plasma had been bound [cf. also LATHEM, 118]. ALLISON and AP REES [4] suggested that this ability to prevent the loss of haemoglobin and thus conserving iron was possibly the main function of haptoglobin [KEENE and JANDL, 111]. Certainly in cases of paroxysmal nocturnal haemoglobinuria and other forms of haemolytic anaemia plasma haptoglobin levels are very low or zero [ALLISON and AP REES, 4; NOSSLIN and NYMAN, 152; BRUS and LEWIS, 45]. This is true for haemoglobinopathies due to Hb S or Hb S and C [MEHTA and JENSEN, 133; WHITTEN, 217, 218; LATHEM and JENSEN, 119]. WHITTEN, however points out that haemoglobin is not always excreted in the urine when haptoglobin levels are low. Studying 58 children with sickle-cell disease and other haemolytic anaemias, all but one of which were ahaptoglobinaemic, he found elevated haemoglobin levels in the plasma and increased excretion of haemosiderin in the urine.

WHITTEN states that free haemoglobin may be filtered by the glomerulus, but that much of it is reabsorbed by the tubular cells. This reabsorbed haemoglobin is degraded by the tubular cells to liberate iron in the form of haemosiderin, some of which may escape in the urine. WHITTEN's observations on the excretion of haemoglobin in patients with sickle-cell disease are in contrast with the observations on haemoglobin excretion in normal subjects referred to earlier. In these cases free haemoglobin appears in the urine as soon as the haemoglobin-binding capacity of the plasma is exhausted. In passing it should be noted that the plasma haptoglobin-level of sickle-cell heterozygotes was not significantly different from that of controls, but there was a significant increase in the number of Hp 0 individuals in the sickle-cell anaemia group [MEHTA and JENSEN, 133].

In normal subjects reabsorption of haemoglobin by the tubules probably accounts for only a small fraction of the total [LATHEM and WORLEY, 120], the clearance of haemoglobin, particularly by the liver, probably playing a major role. Indeed, MURRAY, CONNELL and PERT [141] found that in the nephrectomized rabbit both free Hb and the haptoglobin-haemoglobin complex is removed by the liver, the free haemoglobin being removed more rapidly.

NAKAJIMA [147] and his colleagues have recently established an interesting relationship between haptoglobin group and degradation of haemoglobin in the liver. The enzyme α-haem-methenyl oxygenase catalyses the oxidative cleavage of haem during its degradation to biliverdin. Using animal liver homogenates the Japanese group found that the enzyme is inactive with free haemoglobin, but acts readily on the haptoglobin-haemoglobin complex. When Hb is bound to haptoglobin from different individuals the rate of reaction was essentially the same for serum of groups Hp 1–1 and Hp 2–1 but was significantly greater for group Hp 2–2. SMITHIES [192] has suggested that this difference could be the basis of a selective advantage of the Hp^2 gene.*

Haptoglobin and Disease

Attention has been drawn already to the fact that in a variety of haemolytic diseases, plasma haptoglobin levels are very low, and also that in infections the converse is true. Studies of these changes have been summarized by JAYLE and DORMANN [106], LAURELL and GRONVALL [123] and by OWEN and his colleagues [155, 158] (see Table XI).

The possibility that haptoglobin has some protective effect in preventing the loss of plasma haemoglobin however has stimulated studies on the association between the frequency of the haptoglobin groups and various diseases. In particular such an association might

* Note added in proof: SNELLMAN and SYLVER [Nature 216, 1033 (1967)] have shown that purified haptoglobin inhibits cathepsin activity, and have suggested that it may be important as a protector against proteolysis. They did not test the relative effectiveness of haptoglobin from different Hp groups.

Table XI. Haptoglobin levels in various diseases. (Data from OWEN *et al.* [158])

Condition	No. of cases	Serum haptoglobin level Mean	Standard deviation	Range
Normal	152	93	40	10–220
Infection;				
Acute	17	321	171	62–730
Chronic	12	255	155	74–560
Corrective tissue disorders;				
Rheumatoid arthritis	7	253	69	193–350
Disseminated lupus	6	143	99	15–255
Neoplasia;				
Carcinoma	54	216	93	62–510
Other	64	147	91	10–420
Blood disorders;				
Haemolytic disease	24	17	12	10–48
Megaloblastic anaemia:				
– untreated	25	26	21	10–78
– treated	6	103	–	72–175
Aplastic anaemia	5	168	–	40–343
Hepatobiliary disease;				
Acute hepatitis	13	101	68	10–205
Chronic hepatitis	71	108	100	10–465
Drug hepatitis	11	127	85	11–250
Biliary obstruction	61	190	78	19–410
Renal disease;				
Acute nephritis	3	168	–	125–203
Chronic nephritis	21	173	88	25–400
Chronic pyelonephritis	14	217	115	100–350
Alimentary tract disorders				
Peptic ulcer	19	150	56	80–280
Ulcerative colitis	3	330	–	250–450
Cardiovascular disease;				
Myocardial infarction	40	213	94	58–470
Arterial disease	22	161	69	90–350
Acute rheumatic fever	5	486	–	195–760
Metabolic disease;				
Diabetes mellitus	18	172	63	69–270

be found in situations where haemolytic diseases are prevalent *e.g.* in malarious areas.

In New Guinea, CURTAIN *et al.* [57] have found a significant difference for the frequency of both the Hp^1 gene frequency and for the incidence of Hp 0 individuals between highland and lowland groups. The latter are inhabiting a malarious environment, whilst the former in general are free from the disease. In the Eastern Highlands of New Guinea the Hp^1 frequency is significantly lower than it is in the Markham and Sepik River Valleys; moreover the Hp 0 phenotype was absent in the highlands, but ranged from 2 to 13% in the valley populations.

In addition CURTAIN and his colleagues found significantly lower mean haptoglobin levels in the lowland populations. They suggested that these lowered haptoglobin levels result also in the higher Hp 0 frequency in the lowlands and that the dual effect may be due to increased intravascular haemolysis in the lowland areas. Further since the haptoglobin level in Hp 2–2 and Hp 2–1 persons is lower on the average than in Hp 1–1 persons, loss of haptoglobin as expressed by a higher Hp 0 frequency will tend to increase the apparent frequency of the Hp^1 gene, as found in lowland areas. Alternatively, the authors point out that if there is a selective pressure favouring the Hp^1 gene in lowland areas a similar difference in gene frequency could be established. Since sera from Hp 1–1 and 2–1 persons bind more haemoglobin per 100 ml than do sera from Hp 2–2 persons the Hp 1–1 and 2–1 phenotypes could be at an advantage in maintaining the delicate iron balance in the presence of haemolysis, intestinal iron loss due to parasitic infestation and dietary iron deficiency. This mechanism, of course, would not lead to a new stable equilibrimum in gene frequencies.

The possibility that selective effects are disturbing haptoglobin frequencies has been discussed also by WORKMAN and his colleagues [220] in a study of a Negro population in the United States.

Two other diseases have received some attention in relation to the distribution of haptoglobin phenotypes. ALLISON and BLUMBERG [6] studied the plasma haptoglobin levels and also the distribution of Hp groups in rheumatoid arthritis patients and controls. The levels in patients were elevated above those in the controls, but there was no significant difference in the distribution of the groups. A similar investigation was made by MURRAY and his colleagues [142] for rheumatic fever patients and controls; there was no elevation of plasma haptoglobin level, and no significant difference in the distribution of Hp 1–1, 2–1 and 2–2 between patients and controls, but there was a significant increase in the number of patients with ahaptoglobinaemia (Hp 0). Further when the patients were subdivided there was an increase in the Hp 1–1 phenotype of rheumatoid patients with Sydenham's chorea.

The ahaptoglobinaemic patients were all found in the group with cardiac sequelae. The authours point out that there are several possible explanations for this finding, the most attractive being that persons with low haptoglobin levels may be more susceptible to rheumatic fever leading to complications of this type. NOONAN, HOLLAND and HOLMES [151] however also found 8 of 48 unoperated patients with valvular rheumatic heart disease had low haptoglobin levels and that all 37 patients after insertion of a Starr Edwards valve prosthesis were ahaptoglobinaemic. They believe that valvular disorders lead to mechanical red cell destruction, which in turn leads to a lowering of the haptoglobin level.

Some South American investigations [29] had suggested that patients with leprosy had an abnormal distribution of haptoglobin groups. However a recent large survey [181] of over 1000 leprosy patients in Brazil has failed to reveal any significant differences with the distribution of haptoglobin groups in a large control series.

Finally, IHM and WENDT [98] have reported the haptoglobin group distribution for over 2000 persons in Germany suffering from cardiac infarction, psoriasis, cancer of the colon, cholelithiasis, gastric ulcer and neurodermatitis. No significant differences between any of the disease classes and a control population were noted.

7. WORLD DISTRIBUTION OF THE HAPTOGLOBIN GROUPS

The frequency of the two common alleles Hp^1 and Hp^2 in any population sample can readily be obtained by the method of gene counting. In a population sample of size N, with the proportion of the haptoglobin phenotypes designated by $\overline{1\text{–}1}$, $\overline{2\text{–}1}$ and $\overline{2\text{–}2}$, the gene frequencies are given by

$$Hp^1 = \frac{2\,(\overline{1\text{–}1}) + \overline{2\text{–}1}}{2\text{N}}$$

and

$$Hp^2 = \frac{2\,(\overline{2\text{–}2}) + \overline{2\text{–}1}}{2\text{N}}$$

The standard deviation is the same for both gene frequencies, and is given by the expression

$$\sigma\, Hp^1 = \sigma\, Hp^2 = \sqrt{\frac{\overline{2\text{–}1}}{2\text{N}}}$$

In many populations some haptoglobin phenotypes are scored as Hp 0. Some of these Hp 0 individuals may be genetically deficient in haptoglobin (see Section 4 above), but many are undoubtedly environmental in origin. It is usual therefore to record the Hp 0 frequency but to subtract the number of Hp 0 persons from the total sample size in calculating the gene frequency. Similarly, in populations where the Hp 2–1 (mod) phenotype occurs, the frequency is recorded but the Hp 2–1 and Hp 2–1 (mod) phenotypes are combined for the calculation of the gene frequency. This procedure is justified at present because of both the variable nature of the Hp 2–1 (mod) phenotype [Sutton and Karp, 203] and the lack of certainty about its genetic control.

A. Distribution of the Alleles Hp^1 and Hp^2

The haptoglobin polymorphism appears to be one of considerable antiquity, since it is present in all populations so far studied. However, examination of the results for populations in many parts of the world in Table XIIa and Figure 11 reveals that although the alleles Hp^1 and Hp^2 are everywhere present they vary widely in frequency. At one extreme some African populations have an Hp^1 frequency of 0.7 or higher whilst at the other extreme some Asiatic Indian populations have Hp^1 frequencies as low as 0.1.

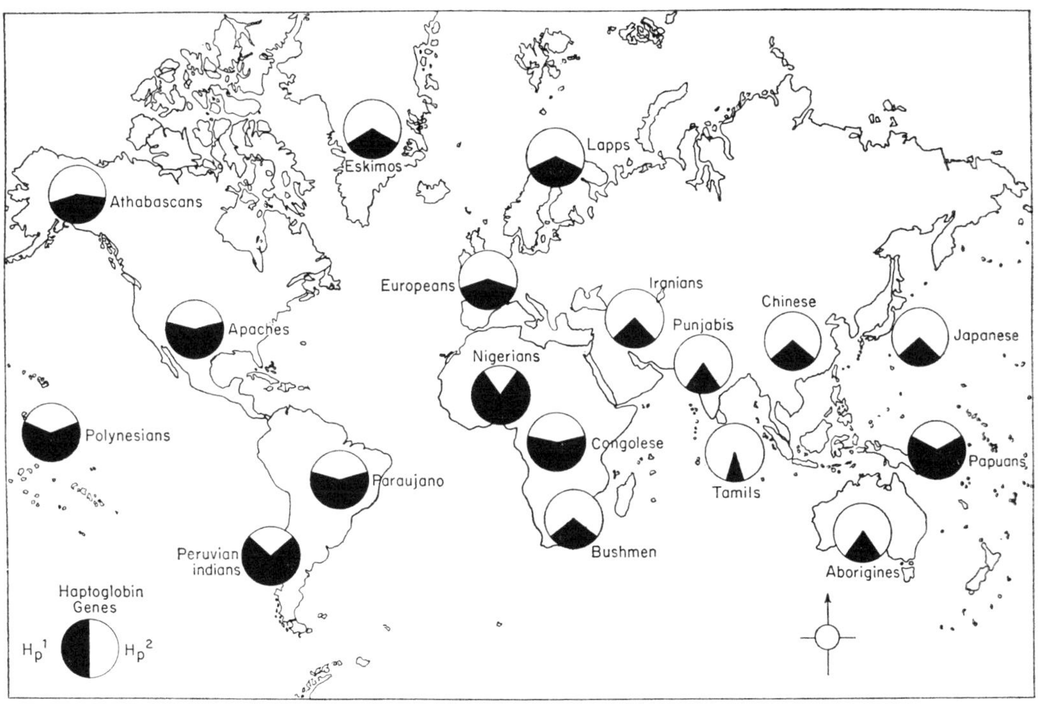

Fig. 11. World distribution of the Hp^1 and Hp^2 genes. Selected populations only.

Table XIIa. Haptoglobin frequencies in selected populations

Notes:
Hp 0 individuals were excluded
– indicates no figure given
* includes 1 Hp 2–1 (Mod) individual
** includes 2 or more Hp 2–1 (Mod) individuals
+ based on gene 'weighting' for related individuals
ø sampled in Malaya

Population	No. tested	Per cent Hp group 1–1	2–1	2–2	0	Hp^1 gene frequency	Ref.
Europe							
Greenland							
(Eskimos)	737	22.9	52.6	23.5	1.0	0.25	162
(Thule Eskimos)	89	11.2	38.2	50.6	0	0.34	80a
Iceland	188	13.3	50.5	35.7	0.5	0.39	215
Iceland	402	18.1	46.5	34.8	0.5	0.42	26
Russia	205	13.7	46.8	38.1	1.4	0.37	1
Poland	2000	12.9	48.5*	38.6	0	0.37	178
Poland	208	11.1	50.0	38.9	0	0.36	139
Poland	3809	14.0	48.0	38.0	–	0.38	116
Finland	891	14.5	41.9	43.4	0.2	0.36	128
Sweden	1003	15.1	44.4	40.1	0.3	0.38	206
Sweden	1279	15.0	48.1	36.3	0.6	0.39	25
Sweden (Lapps)	329	9.4	43.2	45.3	2.1	0.28	28
Sweden (Gypsies)	115	0	23.5	76.5	0	0.12	28a
Norway	5811	14.2	46.8	38.9	–	0.40	64
Denmark	2046	16.0	47.3	36.7	–	0.40	67
England	218	10.1	55.5	31.7	2.7	0.41	8
England	179	18.4	49.1	32.4	–	0.43	91
Belgium	610	18.4	46.2	35.4	0	0.42	209
France	1750	15.5	46.9	37.3	0.4	0.39	138
France	406	15.3	49.7	35.0	0	0.40	136
Spain	2763	16.3	48.3	35.4	–	0.40	163
Spain (Basques)	314	17.5	50.3	30.6	1.5	0.43	163
Portugal	838	15.9	46.6*	37.5	–	0.39	19
Germany	474	19.6	51.1	29.3	0	0.45	13
Austria	621	16.1	48.2	35.7	0	0.40	13
Switzerland	920	15.2	48.8	36.0	–	0.40	47
Czechoslovakia	662	17.1	47.7	35.2	0	0.41	95
Hungary	506	17.0	45.4	37.6	–	0.40	212
Bulgaria	938	10.5	42.0	45.0	2.5	0.34	165
Yugoslavia	459	14.3	47.7	39.8	–	0.37	88
Italy	3714	13.5	46.3	40.3	–	0.37	200
Italy (Milan Province)	596	15.2	44.0	40.3	0.2	0.38	30
Greece	649	9.4	42.1	46.5	2.0	0.31	42
Crete	169	14.8	43.2	41.4	0.6	0.37	18
Cyprus	196	5.1	34.7	57.1	3.1	0.23	163b

Table XIIa. Continuated

Population	No. tested	Per cent Hp group 1–1	2–1	2–2	0	Hp^1 gene frequency	Ref.
Middle East							
Persia (Iran)	97	12.4	46.4	41.2	–	0.35	214
Israel							
(Ashkenazim Jews)	499	8.6	40.5	50.3	0.6	0.29	83
(Oriental Jews)	345	5.8	41.1*	52.2	0.9	0.26	83
(Kurdish Jews)	113	10.6	50.5	37.2	1.7	0.36	170
(Samaritans)	125	16.0	47.2	36.8	0	0.40	42b
India							
Punjabis	161	5.6	30.4	63.4	0.6	0.21	208
Bengalis	176	4.5	29.5*	64.2	1.6	0.20	207
Pathans	185	5.4	36.2	55.7	2.7	0.24	114
Tamils ø	133	3.0	10.5*	83.5	3.0	0.09	114
Oraons	125	1.6	25.6	72.0	0.8	0.15	114
Todas	89	14.6	40.4	43.8	1.1	0.37	114
Irulas	74	0	13.5	79.7	6.8	0.07	114
Ceylon							
Tamils	46	0	28.3	71.7	0	0.14	114
Singhalese	87	0	32.2	65.6	2.3	0.16	114
Veddahs	64	1.6	21.9	70.3	6.2	0.19	114
Thailand							
Thais	413	4.8	37.5	56.4	1.2	0.24	114
Thais	682	5.7	37.1	54.8	2.3	0.24	38
Malaysia							
Malays	236	5.1	36.0*	58.1	0.8	0.23	114
Chinese	167	10.8	34.1	53.9	1.2	0.28	114
Proto-Malays	66	18.2	48.5	24.2	9.1	0.47	114
Philipines							
Filipinos	293	14.0	48.1	37.2	0.7	0.38	36
Taiwan							
Chinese	172	9.3	37.7	52.9	0	0.28	37
Hong Kong							
(Cantonese)	122	13.9	48.4	37.7	0	0.39	177a
Japan							
Japanese	349	6.0	34.9	57.6	1.4	0.24	131
Korea							
South Koreans	120	–	–	–	–	0.32	184

Table XIIa. Continuated

Population	No. tested	Per cent Hp group 1–1	2–1	2–2	0	Hp^1 gene frequency	Ref.
Australia and Oceania							
Australian Whites	323	14.6	46.1	39.0	0.3	0.38	115
Australian Aborigines							
Western Desert	352	3.4	23.9	72.9	0.6	0.15	113
Mitchell River (N. Queensland)	115	2.6	28.7	66.1	2.6	0.17	113
Halls Creek (N. W. Australia)	153	13.1	35.9	49.7	1.3	0.31	113
New Guinea							
Sepik River	623	48.8	37.9	9.1	4.2	0.70	57
Markham River	257	50.2	35.4**	7.8	6.6	0.73	57
Eastern H'lands	705	43.7	43.8	12.5	0	0.67	57
Southern H'lands	222	45.0	42.3	12.6	0	0.66	57
New Britain	821	52.6	37.1	7.9	2.3	0.73	57
Solomon Is.	183	29.0	35.5	8.7	26.8	0.64	58
Tonga Is.	200	36.0	48.5	15.5	0	0.60	59
Truk (Micronesians)	356	14.0	49.4	36.5	–	0.39	163a
Easter Is.	36	72.2	27.8	0	–	0.86+	145
Hawaiian Is.	38	50.7	34.7	14.6	–	0.68	27
Marshall Is.	176	33.5	47.2*	17.6	1.7	0.58	41
North America							
U.S. White	409	13.2	50.3	36.4	0	0.38	77
U.S. Negroes	406	26.4	48.0	21.4	4.2	0.53	73
U.S. Negros	167	28.7	44.3**	25.2	1.8	0.52	54
Eskimos							
(Alaska)	418	7.7	43.5	48.3	0.5	0.32	40
(Alaska)	220	11.8	41.4	46.8	–	0.32	182
Athabascan Indians	284	18.0	31.3	47.9	2.8	0.42	40
Haida Indians	409	27.4	53.5	19.1	–	0.54	127
Apaches Indians	98	34.7	48.0	17.3	–	0.59	205
Nez Perce Indians	180	22.2	48.3	29.4	0	0.48+	65
Navajo Indians	169	23.1	47.3	29.6	0	0.47	159
Central America							
Maya (S. Mexico)	449	36.5	47.7**	14.7	1.1	0.61	204
Maya (Honduras)	212	43.4	44.3	12.3	0	0.66	129
Indians Costa Rica	139	18.0	42.4	39.6	0	0.39	129
Indians Panama	452	25.6	45.6**	25.9	2.7	0.50	129
Mexico (Italians)	150	23.1	48.3	26.5	2.1	0.48	126b
Mexico (Mexicans)	386	22.0	48.4**	24.9	4.7	0.49	126a
Ecuador (Indians)							
Quecha	192	58.3	34.4	4.7	2.6	0.78	129b
Colorado	36	77.8	22.0	0	0	0.89	129b
Jivaro	222	41.4	45.0	12.6	0.9	0.65	129b
Cayapa	226	43.4	44.2	12.4	0	0.67	129b
Secoya	48	29.2	47.9	22.9	0	0.53	129b

Table XIIa. Continuated

Population	No. tested	Per cent Hp group 1–1	2–1	2–2	0	Hp^1 gene frequency	Ref.
South America							
Fr. Guiana (Indians)							
Galibi	178	21.9	58.4**	16.9	2.8	0.53	47a
Palikour	56	50.0	35.7*	10.7	3.6	0.52	47a
Oyampi	95	24.2	43.3**	27.3	5.2	0.48	47a
Emerillon	33	24.2	60.1	12.7	3.0	0.56	47a
Roucouyonne	97	17.5	57.6**	21.0	4.1	0.48	47a
Peru (Indians)							
Piro	86	33.7	47.7	18.6	0	0.58	129a
Campa	93	35.5	39.8	24.7	0	0.55	129a
Shipibo	129	41.9	45.0	13.2	0	0.64	129a
Shipibo	70	35.7	58.6	5.7	–	0.65	35c
Aguaruna	151	19.9	51.7	28.5	0	0.46	129a
Ticuna	122	49.2	36.9	13.9	0	0.68	129a
Aymara	71	50.7	36.6	11.3	1.4	0.69	129a
Quechua	117	54.7	38.5	6.8	–	0.74	76
Venezuela							
Irapa	129	0.8	39.5	59.7	0	0.21	12
Waica	139	61.1	34.6	2.2	2.1	0.80	125
Paraujano	116	28.0	49.0	21.2	1.8	0.53	12
Colombia							
Ica	114	27.2	52.6	15.8	4.4	0.56	70
Paez	103	52.4	41.8	5.8	0	0.73	70
Brazil							
Whites	500	17.0	45.2	36.2	1.6	0.40	181
Aweikoma	112	47.3	36.6*	15.2	0.9	0.65	177
Caingaing	326	53.4	39.4**	7.2	–	0.72	176
Xavante	78	14.1	51.3*	21.8	12.8	0.46	150
Xavante	621	21.9	48.9*	26.7	2.5	0.48	181a
Chile							
Chileans	148	26.3	53.3	20.2	–	0.53	146a
Araucanian Indians	229	59.4	34.1	6.5	–	0.76	144

Populations in Europe have been surveyed in some detail and Hp^1 frequencies range from 0.31 to 0.45, with the majority of values from 0.36 to 0.43. SPEDINI [200] has pointed out that there is a cline across Europe, with Hp^1 values tending to drop as one proceeds eastwards. Though the change is not very great, the lower Hp^1 values in Europe occur in Russia and Poland, but also in Bulgaria, Yugoslavia, Greece and Italy. Extremely low values of Hp^1 occur among Eskimos in Greenland and among Swedish Lapps. The latter, of course, are known to differ markedly from neighbouring Swedish populations for a number of genetic characters.

As one passes from Europe into the Middle East the frequency of the Hp^1 gene falls further and this decline is continued into India, where some of the lowest Hp^1 frequencies in the world are found among the Tamils of S. India. Relatively low values for Hp^1 are found also among the Mongoloid populations of Japan, Thailand, Taiwan and Malaya. Higher values are found among Proto-Malay populations and throughout the islands of the Pacific. By contrast Australian Aborigines have relatively low values.

Indian populations in the Americas show a wide range of Hp^1 gene frequencies. The few results available suggest that the Hp^1 values are lower in N. America, and higher in S. America, with intermediate values in Central America, but the range of frequencies in neighbouring tribes can sometimes be considerable. Thus the Irapa with an Hp^1 value of 0.21 are in marked contrast to another Venezuelan tribe, the Waica with $Hp^1 = 0.80$.

The calculation of haptoglobin gene frequencies in Africa is beset with difficulties. Some of the published results for the observed phenotypes in African populations are shown in Table XIIb. In addition to the three common phenotypes present in other populations, there is often an appreciable frequency of both the Hp 2–1 (mod) and Hp 0 phenotypes present in Africa. The Hp 2–1 (mod) phenotype ranges in occurrence from 0 to 1% in Ethiopia and among the Bushmen to approximately 10% in Senegal and among some Bantu-speaking tribes in South Africa. The majority of reported values, however, fall within the range 2 to 6%. It is interesting to note that the incidence of Hp 2–1 (mod) in American Negroes is approximately 10%, somewhat higher than the majority of values reported from Africa itself.

The reported incidence of the Hp 0 phenotype (approximately 2.0 to 5.0% in American Negroes) is even more variable, ranging from 0 in one study in the Congo to between 30 and 40% in a number of populations in West and Central Africa.

Some of the difference in the reported frequencies for Hp 0 are due undoubtedly to technical difficulties. Under some conditions of electrophoresis free haemoglobin is not readily distinguished from the haptoglobin 1 band. Haemolyzed Hp 0 specimens could therefore be classi-

Table XIIb. Selected african populations. Hp^1 gene frequency calculated on basis of Hp 2–1 (Mod) and 2–1 combined, excluding Hp 0

Note: * Hp 0 probably recorded as Hp 1–1

Population	No. tested	Per cent Hp group 1–1	2–1	2–2	2–1 (Mod)	0	Hp^1 gene frequency	Ref.
Egypt	219	4.1	34.2	61.6	–	–	0.21	92
Ethiopia	312	14.7	45.8	33.0	0.6	5.8	0.40	15
Senegal								
Dakar	396	37.4	35.6	11.9	11.1	4.0	0.61	137
Liberia	429	55.5	37.1	7.4	–	–*	0.74	205
N-W	192	43.8	14.1	9.4	13.0	19.8	0.71	149a
Central	53	37.7	20.8	13.2	9.4	18.9	0.65	149a
S-W	84	40.5	25.0	7.1	7.1	20.2	0.71	149a
Ivory Coast	185	48.1	39.5	12.4	–	–*	0.68	205
Gambia	157	30.6	14.6	7.0	7.0	40.8	0.70	91
North Nigeria								
Habe	120	20.0	43.3	5.8	3.3	27.5	0.60	16
Fulani	111	36.0	19.8	2.7	4.5	36.9	0.76	16
Uganda								
Karamojo	108	20.4	38.0	5.5	2.8	33.3	0.41	2
Baganda	165	28.5	32.7	8.5	6.1	24.2	0.63	5
Burundi								
Tutsi	86	27.9	47.7	22.1	–	2.3	0.52	210
Hutu	96	28.2	47.9	19.8	–	4.1	0.52	210
Congo	99	38.4	44.5	13.1	2.0	2.0	0.61	209
Congo (Leopoldville)	151	59.0	37.5	3.5	–	0*	0.77	199
Congo (Pygmies)	121	6.6	40.5	20.6	0.9	31.4	0.70	78
Bechuanaland Bushmen								
Central	113	10.6	35.4	52.2	0	1.8	0.29	17
Central and Southern	125	9.8	39.3	49.2	1.6	2.3	0.30	110
Bantu	171	31.0	36.3	25.1	4.1	2.9	0.51	110
Central African Republic	353	24.1	22.7	16.7	3.4	33.1	0.56	199a
Republic of South Africa								
Hottentot	59	30.5	42.4	27.1	0	0	0.51	17
Bantu								
Xhosa	265	28.3	36.2	21.5	10.6	3.4	0.54	78
Msutu	218	28.9	38.1	21.1	6.9	5.0	0.54	78
Pedi	78	41.0	24.4	12.8	3.8	17.9	0.55	109
Zulu	113	31.9	39.8	25.7	2.6	2.6	0.53	17
Cape Colored	100	16.0	46.0	38.0	0	0	0.39	84
Madagascar								
Plateau	132	34.8	39.4	25.8	–	–	0.54	35b
Lowland	150	44.0	43.3	12.7	–	–	0.66	35b

fied as Hp 1–1, and this will give a false elevation of the Hp^1 gene frequency.

Where adequate discrimination between the haemoglobin and haptoglobin bands is made, the Hp 0 phenotype will be correctly scored. However, since an unknown proportion of Hp 0 persons have low haptoglobin content of their serum because of environmental factors such as malarial parasitaemia, it is possible only to estimate the Hp^1 and the Hp^2 gene frequencies by omitting the Hp 0 persons from the sample. If the likelihood of becoming Hp 0 were distributed proportionally between the Hp 1–1, 2–1 and 2–2 phenotypes the omission of Hp 0 persons would not affect the gene frequencies. Since, however, it seems likely that Hp 2–2, 2–1 and 2–1 (mod) persons, since they normally have a lower haptoglobin concentration in their plasma than do Hp 1–1 persons are more likely to become Hp 0 because of environmental factors, the omission of these persons from the sample will again falsely elevate the value of Hp^1.

Bearing these facts in mind, we may discount some of the very high values of Hp^1 in Table XIIb. Indeed, all the populations with an apparent Hp^1 gene frequency of 0.63 or higher have a very high Hp 0 incidence or have the possibility of confusing an Hp 0 phenotype as Hp 1–1. Only the Habe of Northern Nigeria among populations with a high Hp 0 incidence have an Hp^1 frequency less than 0.63. The remaining black African populations, excluding the Bushmen and the Karamajo of Uganda, have Hp^1 frequencies ranging from 0.51 to 0.61. The Bushmen are of interest because they show not only low values of Hp^1 but also a low incidence of both Hp 2–1 (mod) and of Hp 0. These characteristics clearly distinguish them from other black African populations.

B. Gene Frequencies for Hp^{1F} and Hp^{1S}

Because of the more tedious procedure involved relatively few population surveys of the haptoglobin groups have included sub-typing. The available information is summarized in Table XIII. In addition to the frequency of the Hp^{1F}, Hp^{1S} and Hp^2 alleles a figure is given for the proportion of Hp^{1S} to all Hp^1 genes.

With the exception of one population sample from Czechoslovakia, European populations have a slightly higher frequency of Hp^{1S} alleles than of Hp^{1F} alleles. The ratio Hp^{1S}/Hp^1 ranges from 0.49 to 0.68.

The Czech population sample, however, with a ratio of only 0.18 is surprisingly different: re-examination is clearly necessary.

The picture which emerges from the limited studies of Mongoloid populations is even more confusing. From the work of SHIM and BEARN [184] it appears that the Hp^{1F} allele is almost completely absent

Table XIII. Gene frequencies for Hp^{1F} and Hp^{1S} in various populations

Note: * Not all individuals sub-typed

Population	No. tested	Hp^{1F}	Hp^{1S}	Ratio Hp^{1S}/Hp^{1}	Reference
Europe					
Germans					
(1) Marburg	232	0.14	0.26	0.65	48
(2) Freiburg	212	0.21	0.20	0.49	112
Greeks	2026	0.14	0.20	0.59	11
Italians	285	0.12	0.25	0.68	184
Czechs	548	0.31	0.07	0.18	112
Middle East					
Iranians	188	0.23	0.10	0.30	112
Asia					
Indians (Bombay)	85	0.05	0.10	0.67	184
Koreans	120	0	0.32	1.00	184
Japanese (1)	170	0.003	0.227	0.98	184
Japanese (2)	134	0.26	0.05	0.16	112
Chinese (New York)	113	0	0.34	1.0	184
Thais	68	0	0.24	1.0	184
Australia					
Aborigines					
(1) N. Queensland	–	0.01	0.17	0.94	66
(2) Arnhem Land	50	0	0.29	1.00	184
(3) W. Desert	101	0.03	0.17	0.85	184
North America					
U.S. Whites	66*	0.13	0.25	0.66	77
U.S. Negroes	222*	0.26	0.29	0.53	77
Nez Perce Indians	180	0.005	0.472	0.99	65
Alabama-Coushatta Indians	143	0	0.37	1.0	184
Eskimo	67	0	0.24	1.0	184
South America					
Chile					
Auracanian Indians	31	0	0.77	1.0	184
Africa					
Nigeria	78	0.47	0.26	0.36	184

in Koreans, Japanese, Chinese, Thais, Eskimos and the Auracanian Indians of Chile. KEHR-LOKE and his colleagues [112], however report a marked excess of Hp^{1F} genes in a sample of 134 Japanese from Tokyo, and in Japanese samples from Hawaii, although the Hp^{1S} allele was in excess of the Hp^{1F} allele, the latter was still significantly more frequent than in any of the Mongoloid populations for which results on sub-typing have been given by other investigators.

American Negroes and black Africans have values not markedly different from the majority of values reported for Europe, although there is a suggestion that the Hp^{1F} allele may be slightly higher in Negroes and Africans than in Europeans.

The Hp^{1F} allele appears to be infrequent or absent in Australian Aborigines. The number investigated, however is small: moreover, no studies have been reported for Melanesians or for Micronesians or Polynesians.

If the technique of sub-typing could be made more reliable, the alleles Hp^{F} and Hp^{1S} could be valuable genetic markers in anthropological studies. At present, however, the widely discrepant results for similar populations reported by different investigators must cast doubt on their usefulness. This is particularly true for the reported frequency of the Hp^{1F} allele in Mongoloid populations, but the wide range of values reported for European populations is equally disturbing. Further investigations, with adequate controls, are needed urgently.

8. REFERENCES

1. Alimova, M. M.: A study on the distribution of haptoglobin types. Vop. med. Khim. *11:* 68–70 (1965).
2. Allbrook, D.; Barnicot, N. A.; Dance, N.; Lawler. S. D.; Marshall, R. and Mungai, J.: Blood groups, haemoglobin and serum factors of the Karamojo. Hum. Biol. *37:* 217–237 (1965).
3. Allison, A. C.: Genetic control of human haptoglobin synthesis. Nature *183:* 1312–1314 (1959).
4. Allison, A. C. and Ap Rees, W.: The binding of haemoglobin by plasma proteins (haptoglobins). Brit. med. J. *2:* 1137–1143 (1957).
5. Allison, A. C. and Barnicot, N. A.: Haptoglobins and transferrins in some East African peoples. Acta genet., Basel *10:* 17–23 (1960).
6. Allison, A. C. and Blumberg, B. S.: The genetically determined serum haptoglobin in rheumatoid arthritis. Arthritis Rheum. *1:* 239–243 (1958).
8. Allison, A. C.; Blumberg, B. S. and Ap Rees, W.: Haptoglobin types in British, Spanish Basque and Nigerian African populations. Nature *181:* 824–825 (1958).
9. Aly, F. W.; Brinker, G.; Deicher, H.; Hartman, F. and Nix, W.: Inherited, immunologically atypical haptoglobin. Nature *194:* 1091–1092 (1962).
10. Aly, F. W.; Brinker, G.; Cleve, H.; Deicher, H.; Hartman, F. and Nix, W.: Die Bestimmung der Haptoglobingruppen mit Hilfe einfacher immunologischer Methoden. Klin. Wschr. *39:* 610 (1961).
11. Angelopoulos, B.; Tsoukantis, A. and Danopoulos, E.: Distribution of haptoglobin subtypes in Greeks. J. med. Gen. *3:* 276–278 (1966).
12. Arends, T. and Gallango, M. L.: Haptoglobin and transferrin groups in Venezuela. Proc. 8th Congr. int. Soc. Blood Trans. Tokyo, 1960, pp. 379–382 (Karger, Basel/New York, 1962).
13. Baitsch, H.; Meier, G.; Schoeller, L. and Kahlich-Koenner, D. M.: Frequencies of the haptoglobin serum groups among the blood donors from Austria and Germany. Nature *186:* 976–977 (1960).
14. Baitsch, H.; Ritter, H.; Goedde, H. W. and Riedel, V.: Formalgenetische Untersuchungen über den Haptoglobinpolymorphismus. Z. Morph. Anthrop. *55:* 175–180 (1964).
15. Barnicot, N. A.; Garlick, J. P.; Adam, A. and Bat-Miriam, M.: A survey of some genetical characters in Ethiopian tribes 111. Haptoglobins and transferrins. Amer. J. phys. Anthropol. N. S. *20:* 175–178 (1962).
16. Barnicot, N. A.; Garlick, J. P. and Roberts, D. F.: Haptoglobin and transferrin inheritance in Northern Nigerians. Ann. hum. Genet. *24:* 171–183 (1960).
17. Barnicot, N. A.; Garlick, J. P.; Singer, R. and Weiner, J. S.: Haptoglobin and transferrin variants in Bushmen and some other South African peoples. Nature *184:* 2042 only (1959).
18. Barnicot, N. A.; Krimbas, C.; McConnell, R. B. and Beaven, G. H.: A genetical survey of Sphakia, Crete. Hum. Biol. *37:* 274–298 (1965).

19. BARROS, F.: Frequence des haptoglobines dans la population du Portugal metropolitain. IXth Congr. Soc. Europ. Haematol., Lisbon, pp. 26–31 (Karger, Basel/New York, 1963).

20. BAYANI-SIOSON, P. S.; LOUCH, J.; SUTTON, H. E.; NEEL, J. V.; HORNE, S. L. and GERSHOWITZ, H.: Quantitative studies on the haptoglobin of apparently healthy adult male twins. Amer. J. hum. Genet. *14:* 210–219 (1962).

22. BEARN, A. G. and FRANKLIN, E. C.: Some genetical implications of physical studies of human haptoglobins. Science *128:* 596–597 (1958).

23. BEARN, A. G. and FRANKLIN, E. C.: Comparative studies on the physical characteristics of the heritable haptoglobin groups of human sera. J. exp. Med. *109:* 55–68 (1959).

24. BECKMAN, L. and GRIVEA, M.: Haptoglobin variations in newborn children. Acta genet., Basel *14:* 159–167 (1964).

25. BECKMAN, L.; HEIKEN, A. and HIRSCHFELD, J.: Frequencies of haptoglobin types in the Swedish population. Hereditas *47:* 599–605 (1961).

26. BECKMAN, L. and JOHANNSSON, E. O.: Haptoglobins and transferrins in the Icelandic population. Acta genet., Basel *17:* 341–344 (1967).

27. BECKMAN, L.; JOHNSON, F. M.; SAKAI, R. K. and WOODS, J. L.: Serum protein variations in Hawaiian population groups. Acta genet., Basel *14:* 309–318 (1964).

28. BECKMAN, L. and MELLBIN, T.: Haptoglobin types in the Swedish Lapps. Acta genet., Basel *9:* 306–309 (1959).

28a. BECKMAN, L.; TAKMAN, J. and ARFORS, K. E.: Distribution of blood and serum groups in a Swedish gypsy population. Acta genet., Basel *15:* 134–139 (1965).

29. BEIGUELMAN, B.: Genética e epidemiologia das doencas transmissiveis com especial referencia a lepra. Ciência e Cultura *17:* 449–460 (1965).

30. BENERECETTI-SANTACHIARA, A. S. and MODIANO, G.: The frequencies of haptoglobin and transferrin types in some villages of the Milan province. Acta genet., Basel *14:* 36–40 (1964).

31. BERGGARD, I. and BEARN, A. G.: Excretion of haptoglobin in normal urine. Nature *195:* 1311–1312 (1962).

32. BERGSTRAND, C. G.; CZAR, B. and TURUKOSKI, P. H.: Serum haptoglobin in infancy Scand. J. clin. lab. Invest. *13:* 576 (1961).

33. BERNINI, L. F.; BORRI-VOLTATTORN, C. and SINISCALCO, M.: Hybridization of haptoglobin sub-units for a better understanding of some rare haptoglobin variants. Paper to 3rd Int. Congr. Hum. Genet., Chicago (1966).

34. BEUING, H.; CLEVE, H. und DEICHER, H.: Immunologische Untersuchungen menschlicher Haptoglobine: Nachweis zweier, vom genetischen Typ unabhängiger Antigendeterminanten. Z. f. Hyg. *151:* 291–310 (1965).

35. BEUING, H.; DEICHER, H.; CLEVE, H.; NIX, W. and ALY, F. W.: Different antigenic determinants of normal human haptoglobins. H. PEETERS ed. Prot. Biol. Fluids, pp. 372–374 (Elsevier, Amsterdam 1964).

35a. BIAS, W. and MIGEON, B. R.: Haptoglobin: A locus on the D_1 chromosome. Amer. J. hum. Genet. *19:* 393–398 (1967).

35b. BUETTNER-JANUSCH, J. and BUETTNER-JANUSCH, V.: Hemoglobins, haptoglobins and transferrins in the peoples of Madagascar. Amer. J. phys. Anthropol. *22:* 163–170 (1964).

35c. BUETTNER-JANUSCH, J.; BOVE, J. R. and YOUNG, N.: Genetic traits and problems of biological parenthood in two Peruvian tribes. Amer. J. phys. Anthropol. *22:* 149–154 (1964).

36. BLACKWELL, R. Q.; CHEN, HSIN-HWA and CHEN, HO-CHIEU: Haptoglobin distribution in a Filipino population. Nature *202:* 814–815 (1964).

37. BLACKWELL, R. Q.; LIN, TSAU-YEN and SHIAO, D. D. F.: Distribution of haptoglobin among Chinese in Taiwan. Nature *193:* 284–285 (1962).

38. Blackwell, R. Q. and Thephusdin, C.: Distribution of haptoglobin among Thais. Nature *197:* 503 (1963).

39. Bloom, G. E.; Gerald, P. S. and Reisman, L. E.: Ring D chromosome: a second case associated with anomalous haptoglobin inheritance. Science *156:* 1746–1748 (1967).

40. Blumberg, B. S.; Allison, A. C. and Garry, B.: The haptoglobins and haemoglobins of Alaskan Eskimos and Indians. Ann. hum. Genet. *23:* 349–356 (1959).

41. Blumberg, B. S. and Gentile, Z.: Haptoglobins and transferrins of two tropical populations. Nature *189:* 897–899 (1961).

41a. Blumberg, B. S.; Kuvin, S. F.; Robinson, J. C.; Teitelbaum, J. W. and Contacos, P. G.: Alterations in haptoglobin levels. J. amer. med. Ass. *184:* 1021–1023 (1963).

42. Blumberg, B. S.; Murray, R. F.; Allison, A. C.; Barnicot, N. A.; Hirschfeld, J. and Krimbas, C.: Serum protein polymorphisms in Greek populations. Ann. hum. Genet. *28:* 189–194 (1964).

42a. Borglin, N. E. and Nyman, N.: Effect of estrogens on the haptoglobin level in the blood. Scand. J. clin. Lab. Invest. *13:* 107 (1961).

42b. Bonné, B.: Genes and phenotypes in the Samaritan isolate. Amer. J. phys. Anthropol. *24:* 1–20 (1966).

43. Boussier, G. H.; Moretti, J. et Jayle, M. F.: Etude de l'haptoglobine. II. Priétes physiques et chimiques des haptoglobines des types 1 et 11 et de leurs complexes avec l'hémoglobine. Bull. Soc. Chim. Biol. *42:* 837–852 (1960).

44. Bridges, C. B.: Salivary chromosome maps. J. Hered. *26:* 60–64 (1935).

45. Brus, I. and Lewis, S. M.: The haptoglobin content of serum in haemolytic anaemia. Brit. J. Haemat. *5:* 348–355 (1959).

46. Burtin, P.; Grabar, P.; Boussier, J. et Jayle, M. F.: Etude immunochimique de l'haptoglobine. Bull. Soc. Chim. Biol. *36:* 1029–1035 (1954).

47. Butler, R.; Metaxas-Buhler, M.; Rosin, S. und Wandrey, R.: Untersuchungen über die Haptoglobin-Gruppen von Smithies. Schweiz. med. Wschr. *89:* 1041–1043 (1959).

47a. Cabannes, R.; Beurrier, A. et Monnet, B.: Etudes des protéines, des haptoglobines et des transferrines chez les Indiens de Guyane Française. Nouv. Rev. Fr. d'Hématol. *5:* 247–260 (1965).

47b. Cheftel, R. I. and Moretti, J.: Sur la structure des haptoglobines humaines. Comp. Rend. Sci. *262:* 1982–1984 (1966).

48. Cleve, H.: Die Verteilung der Haptoglobinuntergruppen in einer Stichprobe gesunder Blutspender aus Hessen. Humangenetik *2:* 115–118 (1966).

49. Cleve, H. und Deicher, H.: Haptoglobin 'Marburg': Untersuchungen über eine seltene erbliche Haptoglobinvariante mit zwei verschiedenen Phenotypen innerhalb einer Familie. Humangenetik *1:* 537–550 (1965).

49a. Cleve, H.; Gorden, S.; Bowman, B. H. and Bearn, A. G.: Comparison of the tryptic peptides and amino acid composition of the Beta polypeptide chains of the three common haptoglobin phenotypes. Amer. J. hum. Genet. *19:* 713–721 (1967).

50. Connell, G. E.; Dixon, G. H. and Smithies, O.: Subdivision of the three common haptoglobin types based on 'hidden' differences. Nature *193:* 505–506 (1962).

51. Connell, G. E. and Shaw, R. W.: The purification of haptoglobin. Canad. J. Biochem. *39:* 1013–1019 (1961).

52. Connell, G. E. and Smithies, O.: Human haptoglobins: Estimation and purification. Biochem. J. *72:* 115–121 (1959).

53. Connell, G. E.; Smithies, O. and Dixon, G. H.: Gene action in the human haptoglobins II. Isolation and physical characterization of alpha polypeptide chains. J. mol. Biol. *21:* 225–229 (1966).

54. Cooper, A. J.; Blumberg, B. S.; Workman, P. L. and McDonough, J. R.: Biochemical polymorphic traits in a U.S. white and Negro population. Amer. J. hum. Genet. *15:* 420–428 (1963).

55. Cooper, D. W.; Lewis, H. B. M. and Nicholls, E. M.: Haptoglobin 'Johnson' in Australian Aborigines. Nature *208:* 694 (1965).
56. Culliford, B. J.: Haptoglobins and transferrins in forensic bloodstains. Nature *198:* 796–797 (1963).
57. Curtain, C. C.; Gajdusek, D. C.; Kidson, C.; Gorman, J. G.; Champness, L. and Rodrigue, R.: Haptoglobins and transferrins in Melanesia. Amer. J. phys. Anthrop. N. S. *23:* 363–379 (1965).
58. Douglas, R.; Jacobs, J.; Hoult, G. E. and Staveley, J. M.: Blood groups, serum genetic factors and haemoglobins in the Western Solomon Islands. Transfusion *2:* 413–418 (1962).
59. Douglas, R. and Staveley, J. M.: Haptoglobins in Tongans. N. Z. med. J. *2:* 391–392 (1960).
60. Eichmann, K.; Deicher, H. and Cleve, H.: Immunologische Analyse der Beziehungen zwischen den drei verschiedenen Typen von Antigendetiminanten normaler menschlicher Haptoglobine. Humangenetik *2:* 271–286 (1966).
61. Falk, H. und Prokop, O.: Die bisherigen Familienuntersuchungen des Berliner Institutes bezüglich der Haptoglobintypen mit weiteren noch unveröffentlichten Familiendaten. Z. ärztl. Fortbild. *55:* 844–846 (1961).
62. Fine, J. M. et Battistini, A.: Etude immunologique des haptoglobines humaines individuelles. Experientia *16:* 57–59 (1960).
63. Fleischer, E. A. and Lundevall, J.: Inheritance of serum groups. Trans. 6th Congr. Europ. Soc. Haematol. *2:* 906–908 (1957).
64. Fleischer, E. A. and Mohr, J.: Concerning the genetics of the human haptoglobin in 126 Norwegian families with 428 children. Acta genet., Basel *12:* 281–291 (1962).
65. Flory, L. L.: Haptoglobin types of the Nez Perce Indians. Nature *197:* 577–578 (1963).
66. Flory, L. L.: Serum factors of Australian Aborigines from N. Queensland. Nature *201:* 508–509 (1964).
67. Galatius-Jensen, F.: On the genetics of the haptoglobins. Acta genet., Basel *8:* 232–247 (1958a).
68. Galatius-Jensen, F.: Rare phenotypes in the Hp system. Acta genet., Basel *8:* 248–255 (1958b).
69. Galatius-Jensen, F.: The haptoglobins—A genetical study. Thesis, Copenhagen, pp. 1–116 (1960).
70. Gallango, M. L. and Arends, T.: Haemoglobin types and blood serum factors in Colombian Indians. Acta genet., Basel *16:* 162–168 (1966).
71. Gerald, P. S.; Warner, S.; Singer, J. D.; Corcoran, P. A. and Umansky, L.: Possible identification of the chromosome bearing the haptoglobin locus. J. clin. Invest. *43:* 1297 (1964).
72. Gerald, P. S.; Warner, S.; Singer, J. D.; Corcoran, P. A. and Umansky, L.: A ring D chromosome and anomalous inheritance of haptoglobin type. J. Pediat. *70:* 172–179 (1967).
73. Giblett, E. R.: Haptoglobin types in American Negroes. Nature *183:* 192–193 (1959).
74. Giblett, E. R.: Variant haptoglobin phenotypes. Cold Spring Harbor Symp. Quant. Biol. *29:* 321–325 (1964).
75. Giblett, E. R.: Recent advances in haptoglobin and transferrin genetics. XIth International Congress of Haematology: Plenary session papers 251–259. Sydney, 1966.
76. Giblett, E. R. and Best, W. R.: Haptoglobin and transferrin types in Peruvian Indians. Nature *192:* 1300–1301 (1961).
77. Giblett, E. R. and Brooks, L. E.: Haptoglobin sub-types in three racial groups. Nature *197:* 576–577 (1963).

78. GIBLETT, E. R.; MOTULSKY, A. G. and FRASER, G. R.: Population genetic studies in the Congo. IV. Haptoglobin and transferrin serum groups in the Congo and in other African populations. Amer. J. hum. Genet. *18:* 553–558 (1966).

79. GIBLETT, E. R. and STEINBERG, A. G.: The inheritance of the serum haptoglobin types in American Negroes: evidence for a third allele Hp^{2M}. Amer. J. hum. Genet. *12:* 160–169 (1960).

80. GIBLETT, E. R.; UCHIDA, I. and BROOKS, L. E.: Two rare haptoglobin phenotypes, 1–B and 2–B, containing a previously undescribed α-polypeptide chain. Amer. J. hum. Genet. *18:* 448–453 (1966).

80a. GILBERG, AA. and PERSSON, I.: Serum protein types in Polar eskimos. Acta genet., Basel *17:* 422–432 (1967).

81. GOEDDE, H. W.; RITTER, H. und WEYRAUCH, U.: Zum Polymorphismus des Haptoglobin: Methodik der Untergruppenbestimmung (subtyping) und Formalgenetik. Humangenetik *1:* 414–422 (1965).

82. GOHLER, W. und BUNDSCHUH, G.: Zwei neue, offenbar genetisch bedingte Haptoglobinphänotypen in einer Sippe. Z. ärztl. Fortb. *56:* 765–767 (1962).

83. GOLDSCHMIDT, E.; BAYANI-SIOSON, P.; SUTTON, H. E.; FRIED, K.; SANDOR, A. and BLOCH, N.: Haptoglobin frequencies in Jewish communities. Ann. hum. Genet. *26:* 39–46 (1962).

83a. GORDEN, S. and BEARN, A. G.: Hemoglobin binding capacity of isolated haptoglobin polypeptide chains. Proc. Soc. exp. Biol. Med. *121:* 846–850 (1966).

84. GORDON, H.; ROBERTSON, M.; BLAIR, J. M. and VOOIJS, M.: Genetic markers in liver disease. S. Afr. med. J. *38:* 734–735 (1964).

85. GOTTLIEB, A. J.; ROSS, J.; GREENBERG, M. and WISCH, N.: Hypohaptoglobinaemia in an American Negro family. Nature *197:* 1214–1215 (1963).

86. GOTTLIEB, A. J.; WISCH, N. and ROSS, J.: Familial hypohaptoglobinemia: a genetically determined trait segregating from G6PD deficiency. Blood *21:* 129-140 (1963)

88. GRUNWALD, P. and HERMANN, C.: Study of several gene frequencies in the Yugoslav population. Nature *199:* 830–831 (1963).

89. GUINAND, S.; TONNELAT, J.; BOUSSIER, G. et JAYLE, M. F.: Propriétés physiques de l'haptoglobine seromucoide α_2 et de sa combinaison hémoglobinique. Bull. Soc. Chim. Biol. *38:* 329–341 (1956).

90. HARRIS, H.; ROBSON, E. B. and SINISCALCO, M.: Atypical segregation of haptoglobin types in man. Nature *182:* 1324–1325 (1958).

91. HARRIS, H.; ROBSON, E. B. and SINISCALCO, M.: Genetics of the Plasma protein variants. In: G. E. W. WOLSTENHOLME and C. M. O'CONNOR, Eds., Biochemistry of Human Genetics, pp. 151–173 (Churchill, London 1959).

92. HASHEM, N.; KAMEL, K. and HAMMOND, E. I.: Haptoglobin phenotypes among Egyptians. J. med. Genet. *3:* 279–280 (1966).

92a. HAUGE, M.: Om blodtypernes anvendelse i den humane genetik. København, pp. 1–203 (1962).

93. HERMAN-BOUSSIER, G.; MORETTI, J. and JAYLE, M. F.: Etude de l'haptoglobine. I. Préparation des haptoglobines humaines des types I et II. Bull. Soc. Chim. Biol. *42:* 817 (1960).

94. HERMAN, E. C.: Serum haptoglobins in hemolytic disorders. J. Lab. clin. Med. *57:* 834–847 (1961).

95. HERZOG, P.; BOHATOVA, J. and BLOCHOVA, L.: Serum proteins with affinity for hemoglobin. II. Haptoglobin types of the population of Prague. Folia Biol. (Praha) *9:* 265–269 (1963).

96. HIGASHI, G. I. and LUBS, H. A.: Quantitative variations of haptoglobin in a Caucasian family. J. med. Genet. *3:* 281–284 (1966).

97. HUSTINX, T. W. J.; BIJLSMA, J. B. and NIJENHUIS, L. E.: Localization of haptoglobin and ABO. Nature *207:* 327–328 (1965).

98. IHM, P. and WENDT, G. G.: Statistische Betrachtung über die Häufigkeit der Haptoglobin-Typen bei verschiedenen Krankheiten. Humangenetik *2:* 186–191 (1966).
99. INGRAM, V. M. and STRATTON, A. O. W.: Human haemoglobin A_2: chemistry, genetics and evolution. Nature *190:* 1079–1084 (1961).
100. JACOB, F. and MONOD, J.: Genetic regulatory mechanisms in the synthesis of proteins. J. Mol. Biol. *3:* 318–356 (1961).
101. JAVID, J.: The nature of the difference between haptoglobin polymers in the phenotypes Hp 2–1 and Hp 2–2. Proc. nat. Acad. Sci. *52:* 663–667 (1964).
102. JAVID, J.: Haptoglobin 2–1 Bellevue, a haptoglobin β-chain mutant. Proc. nat. Acad. Sci. *57:* 920–924 (1967).
103. JAYLE, M. F. et BADIN, J.: Constitution chimique et électrophorese de l'haptoglobine. C. R. Soc. Biol. *144:* 1475–1478 (1950).
104. JAYLE, M. F. et BOUSSIER, G.: Obtention à l'état homogène d'une mucoprotéine sérique, l'haptoglobine. Bull. Soc. Chim. Biol. *36:* 959–969 (1954).
105. JAYLE, M. F.; BOUSSIER, G. et TONNELAT, J.: Isolement à partir du sérum humain de la combinaison hémoglobine-haptoglobine à l'état homogène. Bull. Soc. Chim. Biol. *38:* 343–349 (1956).
106. JAYLE, M. F. et DORMANN, E.: Variations physiopathologiques de l'haptoglobine. In: Les Haptoglobines. Ed. M. F. JAYLE (Masson et Cie., Paris 1962).
108. JAYLE, M. F. and MORETTI, J.: Haptoglobin: Biochemical, genetic and physiological aspects. Progr. Hemat. *3:* 342 (1962).
109. JENKINS, T.: Quoted by TOBIAS, P. V. in The Peoples of Africa south of the Sahara. In: The Biology of Human Adaptability. Eds P. T. BAKER and J. S. WEINER, pp. 111–200 (London, O. U. P. 1966).
110. JENKINS, T. and STEINBERG, A. G.: Some serum protein polymorphisms in Kalahari Bushmen and Bantu: Gamma globulins, haptoglobins and transferrins. Amer. J. hum. Genet. *18:* 399–407 (1966).
111. KEENE, W. R. and JANDL, J. H.: The sites of hemoglobin catabolism. Blood *26:* 705–719 (1965).
112. KEHR-LOKE, E. R.; WEYRAUCH, U.; RITTER, H. und GOEDDE, H. W.: Untersuchungen zur Populationsgenetik der Haptoglobin-Untergruppen. Blut *14:* 97–100 (1966).
113. KIRK, R. L.: Distribution of genetic markers in Australian Aborigines. Occasional papers in Aboriginal Studies No. 4. Canberra, Australian Institute of Aboriginal Studies, pp. 1–67 (1965).
114. KIRK, R. L. and LAI, L. Y. C.: The distribution of haptoglobins and transferrins in South and South East Asia. Acta genet., Basel *11:* 97–105 (1961).
115. KIRK, R. L.; LAI, L. Y. C. and HOGBEN, D. L.: The haptoglobin groups of white Australians. Med. J. Australia *1:* 45–48 (1960).
116. KOBIELA, J.; TUROWSKA, B. and MOSTOWSKI, J.: The Hp and Gm groups in Poland. Vox Sang. *7:* 638–640 (1962).
117. KORNGOLD, L.: Antigenic differences among human haptoglobins. Int. Arch. Allergy *23:* 268–280 (1963).
118. LATHEM, W.: The renal excretion of hemoglobin: regulatory mechanisms and the differential excretion of free and protein-bound hemoglobin. J. clin. Invest. *38:* 652–658 (1959).
119. LATHEM, W. and JENSEN, W. N.: Plasma hemoglobin-binding capacity in sickle-cell disease. Blood *14:* 1047–1056 (1959).
120. LATHEM, W. and WORLEY, W. E.: The distribution of extracorpuscular hemoglobin in circulating plasma. J. clin. Invest. *38:* 474–483 (1959).
121. LAURELL, C. B.: Purification and propeties of different haptoglobins. Clin. Chim. Acta *4:* 79–81 (1959).

122. LAURELL, C. B.: Metal-binding plasma proteins and cation transport. In: The Plasma Proteins, Vol. 1. Ed. F. W. PUTNAM, pp. 349–378 (Academic Press, New York 1960).
123. LAURELL, C. B. and GRONVALL, C.: Haptoglobins. Adv. clin. Chem. *5:* 135–172 (1962).
124. LAURELL, C. B. and NYMAN, M.: Studies on serum haptoglobin level in haemoglobinaemia and its influence on renal excretion of haemoglobin. Blood *12:* 493–506 (1957).
125. LAYRISSE, M. and WILBERT, J.: Indian societies of Venezuela: their blood group types. Caracas, pp. 1–318 (1966).
126. LEWIS, E. B.: Pseudo-allelism and gene evolution. Cold Spring Harbor Symp. Quant. Biol. *16:* 159–174 (1951).
126a. LISKER, R.; LORIA, A. and CORDOVA, M. S.: Studies on several genetic hematological traits of the Mexican population. VIII. Hemoglobin S, Glucose-6-phosphate dehydrogenase deficiency and other characteristics in a malarial region. Amer. J. Hum. Genet. *17:* 179–187 (1965).
126b. LISKER, R.; REYES, G. R.; LOPEZ, G.; PERAL, A. M. and ZARATE, G.: Caracteristicas hematologicas hereditarias de la poblacion Mexicana. X. Estudio de una comunidad de origen italiano. Revist. de Invest. Clin. *18:* 11–21 (1966).
127. LOVETT, C. A.; PRICE, M. and VINCE, B.: Haptoglobin and Gc types in the Haida Indians. Vox Sang. *12:* 151–153 (1967).
128. MÄKELÄ, O.; ERIKSSON, A. W. and LEHTOVAARA, R.: On the inheritance of the haptoglobin serum groups. Acta Genet., Basel *9:* 149–166 (1959).
129. MATSON, G. A.; SUTTON, H. E.; SWANSON, J. and ROBINSON, A. R.: Distribution of haptoglobin, transferrin and hemoglobin types among Indians of Middle America in British Honduras, Costa Rica and Panama. Amer. J. Phys. Anthropol. N. S. *23:* 123–129 (1965).
129a. MATSON, G. A.; SUTTON, H. E.; SWANSON, J. and ROBINSON, A.: Distribution of hereditary blood groups among Indians in South America. ii. In Peru. Amer. J. phys. Anthropol. *24:* 325–350 (1966).
129b. MATSON, G. A., SUTTON, H. E., SWANSON, J.; ROBINSON, A. R. and SANTIANA, A.: Distribution of hereditary blood groups in S. America. i. In Ecuador. Amer. J. phys. Anthropol. *24:* 51–70 (1966).
130. MATSUNAGA, E.: An inert allele Hp^0 at the Hp locus. Jap. J. hum. Genet. *7:* 133–136 (1962).
131. MATSUNAGA, E. and MURAI, K.: Haptoglobin types in a Japanese population. Nature *186:* 320–321 (1960)
132. MATSUNAGA, E.; MURAI, K. and MATSUDA, E.: Inheritance of haptoglobin types in 51 Japanese families. Acta Genet., Basel *12:* 262–280 (1962).
133. MEHTA, S. R. and JENSEN, W. N.: Haptoglobins in haemoglobinopathy: a genetic and clinical study. Brit. J. Haematol. *6:* 250–258 (1960).
134. MORETTI, J.; BOUSSIER, G. et JAYLE, M. F.: Réalisation technique et premières applications de l'électrophorese sur gel d'amidon. Bull Soc. Chim. Biol. *39:* 593–605 (1957).
135. MORTON, N. E.; KRIEGER, H. and MI, M. P.: Natural selection on polymorphisms in N. E. Brazil. Amer. J. hum. Genet. *18:* 153–171 (1966).
136. MOULLEC, J. and FINE, J. M.: Frequencies of the haptoglobin groups in 406 French blood donors. Nature *184:* 196–197 (1959).
137. MOULLEC, J.; FINE, J. M. et LINHARD, J.: Les groupes de l'haptoglobine dans un échantillon de population africaine de Dakar. Rev. Hématol. *15:* 174–182 (1960).
138. MOULLEC, J.; RUFFIÉ, J.; MATTE, CL.; AUDRAN, R. et NOEL, M.: Observations sur la répartition des groupes sériques (haptoglobines, transferrines, groupes Gm) dans quelques populations. Proc. 2nd int. Conf. hum. Genet. *2:* 762–765 (1963).

139. MURAWSKI, K. and MISZCZAK, T.: Haptoglobin types in Poland. Science *133:* 1427 (1961).
140. MURRAY, R. F. and CONNELL, G. E.: Elevation of serum haptoglobin in rabbits in response to experimental inflammation. Nature *186:* 86 (1960).
141. MURRAY, R. F.; CONNELL, G. E. and PERT, J. H.: Role of haptoglobin in clearance and distribution of extracorpuscular haemoglobin. Blood *17:* 45–53 (1961).
142. MURRAY, R. F.; ROBINSON, J. C.; DUBLIN, T. D.; PITT, E. L. and VISNICH, S.: Haptoglobins and rheumatic fever. Brit. med. J. *1:* 762–765 (1966).
143. MURRAY, R. F.; ROBINSON, J. C. and VISNICH, S.: Observations on the inheritance of hypohaptoglobinemia. Acta Genet., Basel *16:* 113–121 (1966).
144. NAGEL, R. and ETCHEVERRY, R.: Types of haptoglobins in Araucanian Indians of Chile. Nature *197:* 187–188 (1963).
145. NAGEL, R.; ETCHEVERRY, R. and GUZMAN, C.: Haptoglobin types in inhabitants of Easter Island. Nature *201:* 216–217 (1964).
146. NAGEL, R. and RANNEY, H. M.: Haptoglobin binding capacity of certain abnormal hemoglobins. Science *144:* 1014–1015 (1964).
146a. NAGEL, R. and SOTO, O.: Haptoglobin types in native Chileans — a hybrid population. Amer. J. phys. Anthropol. *22:* 335–338 (1964).
147. NAKAJIMA, H.; TAKEMURA, T.; NAKAJIMA, O. and YAMAOKA, K.: Studies on heme α-methenyl oxygenase. J. biol. Chem. *238:* 3784–3796 (1963).
148. NANCE, W. E.: Genetic control of hemoglobin synthesis. Science *141:* 123–130 (1963).
149. NANCE, W. E. and SMITHIES, O.: New haptoglobin alleles: a prediction confirmed. Nature *198:* 869–870 (1963).
149a. NEEL, J. V.; ROBINSON, A. R.; ZUELZER, W. W.; LIVINGSTONE, F. B. and SUTTON, H. E.: The frequency of elevations in the A_2 and fetal hemoglobin fractions in the natives of Liberia and adjacent regions, with data on haptoglobin and transferrin types. Amer. J. hum. Genet. *13:* 262–278 (1961).
150. NEEL, J.V.; SALZANO, F.M.; JUNQUERA, P.C.; KEITER, F. and MAYBURY-LEWIS, D.:. Studies on the Xavante Indians of the Brazilian Matto Grosso. Amer. J. hum. Genet. *16:* 52–140 (1964).
151. NOONAN, J. A.; HOLLAND, N. H. and HOLMES, F. C.: Haptoglobin and rheumatic heart disease. Brit. med. J. *1:* 1538 (1966).
152. NOSSLIN, B. F. and NYMAN, M.: Haptoglobin determination in the diagnosis of of haemolytic disease. Lancet *i:* 1000–1001 (1958).
153. NYMAN, M.: Serum haptoglobin: methodological and clinical studies. Scand. J. clin. Lab. Invest. *11:* Suppl. 39, 1–169 (1959).
154. NYMAN, M.: On plasma proteins with heme or hemoglobin binding capacity. Scand. J. clin. Lab. Invest. *12:* 121–130 (1960).
155. OWEN, J. A.; CAREW, J. P.; COWLING, D. C.; HOBAN, J. P. and SMITH, H.: Serum haptoglobins in megaloblastic anaemia. Brit. J. Haematol. *6:* 242–249 (1960).
156. OWEN, J. A.; MACKAY, I. R. and GOT, C.: Serum haptoglobins in hepatobiliary disease. Brit. med. J. *1:* 1454–1457 (1959).
157. OWEN, J. A.; SILBERMAN, H. J. and GOT, C.: Detection of haemoglobin, haemoglobin-haptoglobin complexes and other substances with peroxidase activity after zone electrophoresis. Nature *182:* 1373 (1958).
158. OWEN, J. A.; SMITH, R.; PADANYI; R. and MARTIN, J.: Serum haptoglobin in disease. Clin. Sci. *24:* 1–6 (1964).
159. PARKER, W. C. and BEARN, A. G.: Haptoglobin and transferrin frequencies in an Navajo population: a new transferrin variant. Science *134:* 106–108 (1961).
160. PARKER, W. C. and BEARN, A. G.: Control gene mutations in the human haptoglobin system. Nature *198:* 107–108 (1963a).
161. PARKER, W. C. and BEARN, A. G.: Control gene mutations as a possible explanation of certain haptoglobin phenotypes. Amer. J. hum. Genet. *15:* 159–181 (1963b).

162. PERSSON, I. and TINSGARD, P.: Serum types in East Greenland Eskimos. Acta Genet., Basel *16:* 84–88 (1966).

163. PLANAS, J.; FUSTÉ, M.; VINAS, J. and IRIZAR, J. L.: Haptoglobin types in the Iberian peninsula. Acta Genet., Basel *16:* 371–376 (1966).

163a. PLATO, C. C. and CRUZ, M.: Blood group and haptoglobin frequencies of the Trukese of Micronesia. Acta Genet., Basel *16:* 74–83 (1966).

163b. PLATO, C. C.; RUCKNAGEL, D. L. and GERSHOWITZ, H.: Studies on the distribution of glucose-6-phosphate dehydrogenase deficiency, thalaassemia and other genetic traits in the coastal and mountain villages of Cyprus. Amer. J. hum. Genet. *16:* 267–283 (1964).

164. POLONOWSKI, M. and JAYLE, M. F.: Existence dans le plasma sanguin d'une substance activant l'action peroxydasique de l'hémoglobine. C. R. Soc. Biol. *129:* 457–460 (1938).

165. POPWASSILEW, I.: Über die Verteilung der Haptoglobintypen in Sofia. Z. ärztl. Fortbildg. *56:* 770 (1962).

166. PROKOP, O.; BUNDSCHUH, G. und FALK, H.: Neue Ergebnisse auf dem Gebiete der Haptoglobine. Dtsch. Z. ges. Med. *51:* 480–487 (1961).

167. PROKOP, O.; FUNFHAUSEN, G. und RACHWITZ, A.: Familienuntersuchungen zur Frage der Vererbung der Eigenschaft Gm^a (gleichzeitig mit Hp-Befunden). Blut *7:* 301–302 (1961).

168. PROKOP, O. und UHLENBRUCK, G.: Lehrbuch der menschlichen Blut- und Serumgruppen. 2. Aufl. (G. Thieme, Leipzig 1966).

169. QUEEN, K. G. and PEACOCK, A. C.: Haptoglobin type determination using noncarcinogenic reactions. Results on 192 normal subjects in Maryland. Clin. Chim. Acta *13:* 47–51 (1966).

170. RAMOT, B.; DUVDEVANI-ZIKERT, P. and KENDE, G.: Haptoglobin and transferrin types in Israel. Ann. hum. Genet. *25:* 267–272 (1962).

171. RAMOT, B.; KENDE, G. and ARNON, A.: Johnson type haptoglobin. Nature *196:* 176 (1962).

172. REICH, M.: Interaction of haemoglobin with serum proteins as shown by electrophoresis. Austr. J. exp. biol. med. Sci. *34:* 151–156 (1956).

173. RENWICK, J. H. and MARSHALL, H.: A new type of human haptoglobin, Hp 2–1D. Ann. hum. Genet. *29:* 389–390 (1966).

174. ROBERT, L.; BOUSSIER, G. et JAYLE, M. F.: Groupements sulfhydriques de l'haptoglobine et de sa combinaison hémoglobinique. Experientia *13:* 111–112 (1957).

175. ROBSON, E. B.; GLENN-BOTT, A. M.; CLEGHORN, T. E. and HARRIS, H.: Some rare haptoglobin types. Ann. hum. Genet. *28:* 77–86 (1964).

176. SALZANO, F. M. and SUTTON, H. E.: Haptoglobin and transferrin types in Southern Brazilian Indians. Acta Genet., Basel *13:* 1–8 (1963).

177. SALZANO, F. M. and SUTTON, H. E.: Haptoglobin and transferrin types of Indians from Santa Catarina, Brazil. Amer. J. hum. Genet. *17:* 280–289 (1965).

177a. SANFORD, R.; GRIMMO, A. E. P. and LEE, SHUN-KEUNG: The haptoglobin and transferrin types of some Cantonese in Hong Kong. Vox Sang. *11:* 106–107 (1966).

178. SCHLESINGER, D.: The haptoglobin group system: Distribution and heredity in the Polish population. Arch. Imm. Ter. Dosw. *9:* 809–846 (1961).

179. SCHLESINGER, D. and MIKULEWICZ, W.: Haptoglobin of the Johnson type in a polish family. Arch. Imm. Ter. Dosw. *12:* 561–564 (1964).

180. SCHULTZE, H. E., HAUPT, E. H., HEIDE, K. und HEIMBURGER, N.: Über die Chemie des Haptoglobin-Polymorphismus. Clin. Chim. Acta *8:* 207–214 (1963).

181. SCHWANTES, A. R.; SALZANO, F. M.; DE CASTRO, I. V. and TONDO, C. V.: Haptoglobins and leprosy. Acta Genet., Basel *17:* 127–136 (1967).

181a. SHREFFLER, D. C. and STEINBERG, A. G.: Further studies on the Xavante Indians. IV. Serum protein groups and the SC_1 trait of saliva in the Simoes Lopes and Sao Marcos Xavantes, Amer. J. hum. Genet. *19:* 514–523 (1967).

182. SCOTT, E. M.; DUNCAN, I. W.; EKSTRAND, V. and WRIGHT, R. C.: Frequency of the polymorphic types of red cell enzymes and serum factors in Alaskan Eskimos and Indians. Amer. J. hum. Genet. *18:* 408–411 (1966).

183. SEREBROVSKY, A. S.: A general scheme for the origin of mutations. Amer. Nat. *63:* 374–378 (1929).

184. SHIM, BONG-SOP and BEARN, A. G.: The distribution of haptoglobin subtypes in various populations, including subtype patterns in some non-human primates. Amer. J. hum. Genet. *16:* 477–483 (1964a).

185. SHIM, BONG-SOP and BEARN, A. G.: Immunological and biochemical studies on serum haptoglobin. J. exp. Med. *120:* 611–628 (1964b).

186. SHIM, BONG-SOP; LEE, TONG-HO and KANG, YOOK-SE: Immunological and biochemical investigations of human serum haptoglobin: composition of haptoglobin-haemoglobin intermediates, haemoglobin binding sites and presence of additional alleles for the β-chain. Nature *207:* 1264–1267 (1965).

187. SMITH, H.; EDMAN, P. and OWEN, J. A.: N-terminal amino acids of human haptoglobins. Nature *193:* 286–287 (1962).

188. SMITHIES, O.: Grouped variations in the occurrence of new protein components in normal human serum. Nature *175:* 307–308 (1955a).

189. SMITHIES, O.: Zone electrophoresis in starch gels: group variations in the serum proteins of normal human adults. Biochem. J. *61:* 629–641 (1955b)

190. SMITHIES, O.: Zone electrophoresis in starch gels and its application to studies of serum proteins. Adv. Prot. Chem. *14:* 65–113 (1959a).

191. SMITHIES, O.: An improved procedure for starch gel electrophoresis. Biochem. J. *71:* 585–587 (1959b).

192. SMITHIES, O.: Chromosomal rearrangements and protein structure. Cold. Spring. Harbor. Symp. Quant. Biol. *29:* 309–319 (1964).

193. SMITHIES, O. and CONNELL, G. E.: Genetic and chemical aspects of the haptoglobins. 2nd int. Conf. hum. Genet. Rome, Proceedings, p. 180 (1961).

194. SMITHIES, O.; CONNELL, G. E. and DIXON, G. H.: Chromosomal rearrangements and the evolution of haptoglobin genes. Nature *196:* 232–236 (1962a).

195. SMITHIES, O.; CONNELL, G. E. and DIXON, G. H.: Inheritance of haptoglobin subtypes. Amer. J. Hum. Genet. *14:* 14–21 (1962b).

196. SMITHIES, O.; CONNELL, G. E. and DIXON, G. H.: Gene action in the human haptoglobins. I. Dissociation into constituent polypeptide chains. J. Mol. Biol. *21:* 213–224 (1966).

197. SMITHIES, O. and WALKER, N. F.: Genetic control of some serum proteins in normal humans. Nature *176:* 1265–1266 (1955).

198. SMITHIES, O. and WALKER, N. F.: Notation for serum protein groups and the genes controlling their inheritance. Nature *178:* 694–695 (1956).

199. SONNET, J. and MICHAUX, J. L.: Glucose-6-phosphate dehydrogenase, haptoglobin groups, blood groups and sickle-cell trait in the Bantus of West Belgian Congo. Nature *188:* 504–505 (1960).

199a. SPEDINI, G.: Le aptoglobine sieriche in 353 negri N'Zakara (R. C.A.) Riv. Antrop. *68:* 145–154 (1961).

200. SPENDINI, G.: Le aptoglobine in antropologia. Acta Genet. Med. Gemell. *13:* 388–399 (1964).

201. STEINBERG, A. G.; LAI, L. Y. C.; VOS, G. H.; SINGH, R. B. and LIM, T. W.: Genetic and population studies of the blood types and serum factors among Indians and Chinese from Malaya. Amer. J. hum. Genet. *13:* 355-371 (1961).

202. STERN, C.: Principles of Human Genetics. 2nd Ed., pp. 1–753 (Freeman, San Francisco 1960).

203. SUTTON, H. E. and KARP, G. W.: Variations in heterozygous expression at the haptoglobin locus. Amer. J. hum. Genet. *16:* 419–434 (1964).

204. Sutton, H. E.; Matson, G. A.; Robinson, A. R. and Koucky, R. W.: Distribution of haptoglobin, transferrin and hemoglobin types among Indians of Southern Mexico and Guatemala. Amer. J. hum. Genet. *12:* 338–347 (1960).

205. Sutton, H. E.; Neel, J. V.; Livingstone, F. B.; Binson, G.; Kunstadter, P. and Trombley, L. E.: The frequencies of haptoglobin types in five populations. Ann. hum. Genet. *23:* 175–183 (1959).

206. Tarukoski, P. H.: Gene frequencies of haptoglobin types in a Swedish population. Nord. Med. *62:* 1425–1426 (1959).

207. Tiwari, S. C.: Haptoglobin and transferrin variants in some upper castes of Bengal. Anthropologist *7:* 17–20 (1960).

208. Tiwari, S. C.: The frequencies of serum haptoglobins and transferrin types in a Punjabi population. Anthropologist *8:* 92–94 (1961).

209. Van Ros, C.; Van Sande, M. et Druet, R.: Groupes d'haptoglobine et de transferrine dans des populations africaines et européennes. Ann. Soc. Belge Med. Trop. *43:* 511–534 (1963)

210. Van Sande, M.; Van Ros, G. and Druet, R.: Determination of haptoglobin groups frequencies by starch gel and agar gel electrophoresis: application to Belgian and Barundi populations. Nature *197:* 603–604 (1963).

211. Veerts, G.; Nix, W. and Deicher, H.: Isolierung und nähere Charakterisierung eines neuen Haptoglobins: Hp-Marburg. Blut *12:* 65–77 (1965).

212. Walter, H.: Untersuchungen zur Häufigkeitsverteilung der Serumgruppen Hp, Gc und Gm in Ungarn sowie in West- und Osteuropa. Anthrop. Anz. *29:* 313–321 (1965).

213. Walter, H.; Bernhard, W. und Heyde, G.: Beitrag zur formalen Genetik der Hp- und Gc-Faktoren: Untersuchungen an 56 Familien mit 182 Kindern. Z. Menschl. Vererb. Konst. *37:* 118–121 (1963).

214. Walter, H. und Djahanschahi, D.: Zur Häufigkeit der Serumgruppen in Persien. Homo *14:* 70–76 (1963).

215. Walter, H. and Palsson, J.: On the distribution of haptoglobin types in Iceland. Amer. J. Phys. Anthropol. N. S. *21:* 49–52 (1963).

216. Wendt, G. G.: Serumeiweiss und Vaterschaft. Dtsch. Med. Wschr. *88:* 916–925 (1963).

217. Whitten, C. F.: Studies on serum haptoglobins. Amer. J. Dis. Child. *102:* 480–481 (1961).

218. Whitten, C. F.: Studies on serum haptoglobin — a functional inquiry. New Engl. J. Med. *266:* 529–534 (1962).

219. Winterhalter, K. H. and Huens, E. R.: Preparation, properties and specific recombination of $\alpha\beta$ globin subunits. J. Biol. Chem. *239:* 3699–3705 (1964).

220. Workman, F. L.; Blumberg, B. S. and Cooper, A. J.: Selection, gene migration and polymorphic stability in a U.S. white and Negro population. Amer. J. hum. Genet. *15:* 429–437 (1963).